Designing Competitive FTC Robots

From Strategy to Robot Excellence

Denny C. Davis, PhD

Copyright © 2021 by Denny C. Davis

All rights reserved

Independently published through Kindle-Amazon
in the United States of America
Available for purchase through Amazon.com

Other books by the author available through Amazon.com:
Teamwork Minutes
Design Thinking
Professional Teamwork Mentor
Project Design Reviews
Hope Beyond High-Risk Multiple Myeloma
Pre-Engineering Primer, 1st Edition and 2nd Edition

Author contact:
Denny C. Davis, PhD
Verity Design Learning
http://veritydesignlearning.com
DennyCDavis73@gmail.com

Acknowledgements

I would like to express my thanks to the many faculty members and students in my life as an engineering design faculty member at Washington State University for over 35 years. Thanks also to engineering design education faculty across the nation who deepened my understanding of design learning and assessment. Through our interactions, you have taught me, encouraged me, and challenged me to think more deeply about how design is done.

Thank you to the team members and coaches of Robo Raiders FTC 7129 who have worked alongside me as we learned to apply engineering design to competition robots. Coaching the Robo Raiders has given me a laboratory for learning how students new to design can apply engineering principles and processes to robot design.

Thanks go to the FIRST® (For Inspiration and Recognition of Science and Technology) organization that inspires young people to engage in robot and professional development through their programs. FIRST® FTC provides a context for learning engineering through robot design.

A big thank you goes to Ken Pugsley, coach of FTC 3409 Astromechs, for his professional review, insightful feedback, and helpful suggestions on the book.

I want to express my special thanks to my wife Irma, who has been by my side for half a century, encouraging me in my work adventures. Her encouragement has not waned through my many experiments in learning and life.

I thank God, the Creator of all things, for enabling me with abilities to do design and for the opportunities to teach design in college and high school contexts. Thanks go to Jesus Christ who freed me from fear of failure and gave me an excitement to use the God-given gifts I enjoy. May this book build abilities of young people and inspire them to enter engineering careers where they can contribute good to society.

Denny C. Davis, PhD

How to Use this Book

This book is written for FTC (FIRST® Tech Challenge) teams creating competition robots. FTC robots must be viable for multiple competitions and achieve greater excellence with each succeeding competition. The book presents design principles and draws examples from robot development with the FTC 7129 Robo Raiders team.

The book has two sections: Robot Design and FTC Robot Development Journey. Chapters 1 through 5 (Section 1) explain the engineering design process. In Section 2, Chapters 6 and 7 show how to do stepwise development of FTC robots, and Chapter 8 shows how to communicate your robot development journey.

Chapter 1 is a cursory walk-through of the design process. It explains four major stages of design. It ends with a mini project to practice the four design stages.

Chapters 2 through 5 more fully explain the engineering design process. Within the four stages of design, twelve design steps are discussed with examples. Chapter 2 takes teams through the Problem Definition stage. Chapter 3 discusses the Conceptual Design stage. Chapter 4 addresses the Prototyped Solution stage. Chapter 5 discusses the Solution Completion stage. Each chapter contains review questions with answers. Suggestions are given for documenting each step. Design reviews are prescribed at the end of each design stage. These chapters show once-through design of a robot, but teams should expect to revisit previous steps (use design iteration) many times.

The FTC season calls for somewhat constrained application of the design process. Chapter 6 walks FTC teams through rapid application of the process when preparing for a first competition only weeks away. It helps teams set reasonable goals and focus effort to create a minimum viable robot for their first competition.

Chapter 7 leads FTC teams through planning and completing robot improvements after one competition and prior to the next. It helps teams review accomplishments and set priorities for pursuing robot excellence for the next competition and longer-term.

Chapter 8 offers teams suggestions and examples for communicating their robot development journey as might be done in an engineering portfolio.

How should you use this book? Use Chapter 1 to lay a foundation for understanding design. Then use Chapters 2 through 5 to study engineering design in detail prior to designing a robot. Use Chapter 6 at the start of an FTC season once the robot game is known. Use Chapter 7 to define and execute robot improvement between competitions. Chapter 8 will help you share highlights of your robot development journey and realize excellence in both your robot and your team.

Table of Contents

Acknowledgements .... *iii*

How to Use this Book .... *iv*

SECTION 1: ROBOT DESIGN 1

Chapter 1: Introduction To Robot Design 2
    Robot Excellence 3
    What is Engineering Design? 4
    Stages of Engineering Design 5
    Mini-Project: Creating Value from a Waste Material 14

Chapter 2: Problem Definition (Stage 1) 15
    Step 1: Defining Game Strategy 16
    Step 2: Defining Robot Needs 21
    Design Review: Problem Definition Stage 25

Chapter 3: Conceptual Design (Stage 2) 27
    Step 3: Generating Component Ideas 28
    Step 4: Screening Component Ideas 32
    Step 5: Creating Robot Concepts 36
    Step 6: Selecting Robot Concept 39
    Design Review: Conceptual Design Stage 43

Chapter 4: Prototyped Solution (Stage 3) 45
    Step 7: Specifying Solution Requirements 46
    Step 8: Building Solution Prototype 50
    Step 9: Testing Solution Prototype 57
    Design Review: Prototyped Solution Stage 61

Chapter 5: Solution Completion (Stage 4) 63
    Step 10: Defining Robot Details 64
    Step 11: Assembling Completed Robot 70
    Step 12: Evaluating Completed Robot 74
    Design Review: Solution Completion Stage 83

SECTION 2: FTC ROBOT DEVELOPMENT JOURNEY 85

Chapter 6: Designing a Minimum Viable Robot 86

- Week 1, Day 1: Defining Game Strategy, Minimum Viable Needs 90
- Week 1, Day 2: Generating Component Ideas 92
- Week 1, Day 3: Screening Component Ideas 94
- Week 1, Day 4: Creating Robot Concepts 95
- Week 1, Day 5: Selecting Robot Concept, Conceptual Design Review 96
- Weeks 2 to 6: Building and Testing Prototype MVP Robot 98
- Weeks 7 to 8: Preparing and Testing Competition MVP Robot 101

Chapter 7: Designing a Robot of Excellence 105

- Defining Game Strategy 106
- Defining Robot Needs 108
- Generating Component Ideas 111
- Screening Component Ideas 112
- Creating System Concepts 113
- Selecting Robot Concept 114
- Specifying Solution Requirements 115
- Building Solution Prototype 116
- Testing Solution Prototype 117
- Defining Robot Details 118
- Assembling Completed Robot 120
- Evaluating Completed Robot 121
- Iteration and Pursuit of Excellence 122

Chapter 8: Communicating Your Pursuit for Excellence 124

- Robot Journey Introduction 125
- Robot Journey Body 126
- Robot Journey Closing 129
- Final Thoughts 130

SECTION 1: ROBOT DESIGN

This section discusses the engineering design process as it relates to robot design. Robot design has four major stages, comprised of twelve design steps. This section focuses on developing understanding of how design activities, some of which are creative and others judgmental, produce excellence in robot design.

    Chapter 1 provides an overview of the robot design process.

    Chapters 2 through 5 detail the steps in robot design.

Whoever loves instruction loves knowledge.

Chapter 1: Introduction To Robot Design

Go, team! Design an awesome robot! This challenge is heard each year by FIRST® Tech Challenge (FTC) and other high school robotics teams around the world. To the novice, designing a robot sounds simple, but people knowledgeable about design know that engineering design is a complex process requiring both creativity and judgment, done well . . . and at the proper time.

This chapter lays a foundation for doing design. It defines design and describes four major stages of robot design. This overview shows that simple engineering design is applicable to the creation of a solution to address a challenge or an opportunity, such as faced by high school robotics teams.

Extensive discussions of design steps in Chapters 2 through 5 show how design can be done as a continuum with attention to achieving both strong robot performance and design excellence. Chapters 6 and 7 present ways to apply the design process when interrupted by multiple competitions during a season.

Be encouraged! You can do engineering design with the degree of sophistication that fits your design team and your design objectives.

Design is a team effort. Design is also an activity that helps develop both technical and professional skills in participants. And . . . as you engage in design, you will begin to think and work like an engineer.

Get onboard for an exciting engineering design journey!

Robot Excellence

Imagine this:

Your robot is the talk of the robotics community! It is simple, attractive, and solidly built. Its innovative features "wow" other teams, coaches, and judges. Your robot travels on the field with quick, smooth, and precise movements. Mechanisms are agile, stable, and exacting. Your drive team can easily make your robot out-perform other robots and score consistently even under pressure. Posted videos of your robot have gone viral, and teams around the world are asking about your amazing design features. Your team has a reputation for robot design excellence.

Does this sound preposterous? Maybe. This level of excellence does occur, but infrequently and not by accident. It requires excellence in all aspects of design. First, you must fully understand the game challenge. You must choose and develop your very best ideas into robot concepts that produce exceptional functionality. Incorporate innovation into your design to improve performance and make your robot memorable. Give attention to selecting excellent materials and parts and using high-quality manufacturing processes so your robot operates smoothly and is durable. Use sensors and controls to improve your robot's actions, simplify operator effort, and save precious time in matches. Seek to integrate form, function, and appearance in your robot.

Your design team is key to achieving excellence in robot design. Your understanding of engineering, physics, manufacturing processes, and teamwork must grow throughout the season and aid you as you make design decisions, execute them, and explain them to others. Applying your knowledge, giving attention to detail, and persisting through challenges will produce a world-class robot . . . and you will become more like engineers as you practice engineering design.

What is Engineering Design?

What is meant by engineering design? Engineering design is the process of creating something (a product, in this case a robot) that brings to life a needed technological solution to a problem or to address an opportunity. For you to create a competitive robot requires that your envisioned robot addresses the game challenge, does not exceed your team's capabilities, and fits time and resources available to your team.

Is design an individual or team sport? Design is best done by a team because it requires creativity, judgment, and other diverse abilities and ways of thinking. Team members, each with different ways of thinking, complement one another and produce designs that are both innovative and effective. However, your members must learn to work as a team, so members' contributions build excellence rather than divide the team.

Robot design is a process that begins with a design challenge and ends with a successfully performing design solution, a robot, as shown in Figure 1.1. The process moves left to right, applying creativity and judgment from the challenge moving toward the envisioned solution, probably with many detours, failures, and successes enroute. The design process is discussed in much detail beginning in Chapter 2.

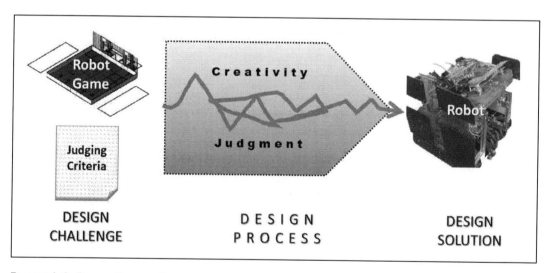

FIGURE 1.1. ROBOT DESIGN PROCESS

Stages of Engineering Design

Engineering design progresses through four major stages: Problem Definition, Conceptual Design, Prototyped Solution, and Solution Completion, as shown in Figure 1.2. Work during each stage produces a corresponding product: Defined Problem, Solution Concept, Solution Prototype, and Completed Solution, respectively.

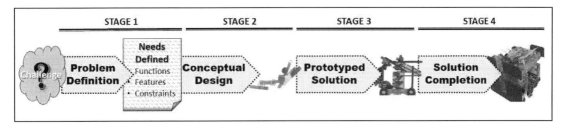

FIGURE 1.2. FOUR MAJOR STAGES OF ENGINEERING DESIGN

Each intermediate product must be completed with excellence if a final design is to be excellent. In fact, your design team should not advance to the next stage of the design process until you have completed the previous stage very well.

This suggests that you hold a design review conducted by an external reviewer at the end of each stage to ensure that your work is done well. A good design review will identify weaknesses that must be addressed before proceeding. Catching weaknesses early in the design process is essential to prevent wasted effort and to produce a high-quality design. For more information on design reviews, see the following video produced as part of a series for FTC teams drawing on the advice of relevant experts:

    Design Reviews   *https://youtu.be/NlI1RCR_3Rk.*

Each of the four stages of the design process is discussed briefly below.

Problem Definition Stage

Begin Problem Definition as soon as the design challenge (robot game) is known. To define the problem, your team defines what your robot must do and be for successful matches and season. First, decide how you will play the game, and define what feats your robot must achieve— types of maneuvers, speed and control required, and physical limitations (size, weight, etc.) Study game rules to be sure you fully understood what is allowed. Remember to identify any desired features that will make your robot stand out in function or appearance.

Your Defined Problem might be simple for your first robot competition, but it needs to be more ambitious for later, highly competitive events. The Defined Problem typically includes two parts: a goal statement and a list of specific needs to be addressed.

A sample Defined Problem for a robot is presented in Figure 1.3. The goal statement describes what the robot is expected to do in each portion of the robot competition. Needs are listed with descriptions and numbers when these are known. Together the goal and list of needs communicates what the robot must do to fulfill the team's game strategy.

Robo Goal: In autonomous, the robot will deliver a wobble goal to the proper target zone and then park on the launch line. Under driver control, it will launch rings into the high goal. In end game, it will deliver a wobble goal to the drop zone and continue scoring rings in the high goal.

Robot Needs: The robot must do the following:
- Fit within 18-inch cube at start of match
- Travel to designated location ±6 inches under program control
- Sense the number of rings 3 feet away
- Hold a wobble goal for transport, 12-inch lift, and release
- Collect 3 rings from the floor and index them for launching
- Launch rings 5.5 feet with horizontal orientation, within 15-inch width and 6-inch height range

---

FIGURE 1.3. SAMPLE DEFINED PROBLEM FOR A COMPETITION ROBOT

Conceptual Design Stage

You will begin Conceptual Design after a clear and complete Defined Problem is prepared. Your robot concept will be a set of features or systems that work together to accomplish your robot goal. Typically, your envisioned robot will have multiple systems, each of which can satisfy a set of needs, and when integrated into an entire robot, meet the defined goal.

As your team examines the list of needs defined above, identify necessary robot components or systems (drive train, collection system, launcher, wobble goal grabber, etc.) to meet these needs. For each necessary system, search for ideas of types that could possibly meet the stated needs. For example, if the game requires gathering rings distributed on the field, look for types of mechanisms that can collect rings off the floor.

Search widely for good ideas. Look at ways others have done similar things. Brainstorm for additional ideas. Compile a list of feasible and possibly creative ideas for each system. Strive to identify several ideas, hopefully including novel ideas, for each system.

Next your team should create robot concepts that incorporate good ideas for each system. As shown in Figure 1.4, multiple ideas for each system will yield several possible robot concepts to consider. For example, ideas for a drivetrain are combined with ideas for a collection system and ideas for a launcher to define feasible concepts for a ring scoring system. These are combined with ideas for other systems to define possible robot concepts.

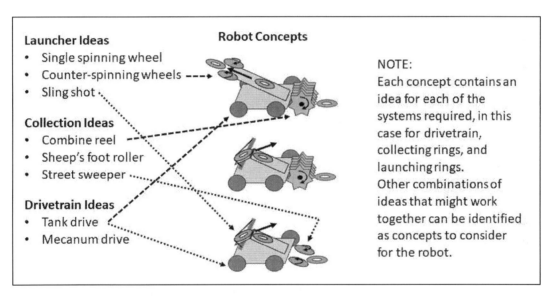

FIGURE 1.4. DEFINING CONCEPTS FROM IDEAS

After defining multiple feasible concepts for the robot, your team should select the best one. You might need to test models, such as physical or CAD (computer-aided drawing) or math models, to determine how well each concept will work under game conditions. Discuss the potential of each concept option as well as the capabilities of your team to bring it to completion before you select the best concept. You might ask external reviewers to help you judge functionality, durability, and manufacturability of your selected concept before the selection is finalized.

Your robot solution concept will be general but define key features of what the team plans to build and program for competitions. Your concept definition should show how desired functionality is to be achieved and should describe key features of major parts. Figure 1.5 shows a sketch of a conceptual design for a ring collection, transfer, and launching system, with key features labelled. A full robot conceptual design would include additional systems to address other scoring to be attempted in the game.

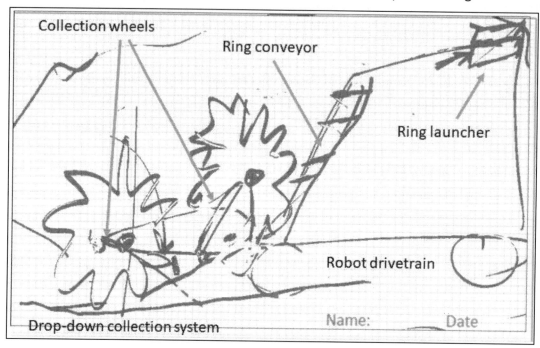

FIGURE 1. 5. CONCEPTUAL DESIGN OF RING COLLECTION AND SCORING SYSTEM

When you have selected your robot concept, you will have tentatively committed to building and programming a robot with its designated types of systems. But before building the entire robot based on this concept, the next stage of the design process will guide you to verify that the concept will perform as expected.

Prototyped Solution Stage

The third stage of design is to prototype and test the conceptual design, with an eye toward its development into the competition robot desired. First you define specific requirements that the conceptual design must achieve to be successful. Here you define how (how fast, how far, how accurately, etc.) the various systems in the concept must perform. You define a set of solution requirements to be met by the prototype.

You then build a prototype of the solution, building only parts needed for testing to see that requirements are achieved. For example, if you defined a requirement that the robot launch rings 5 feet to a target 30 inches from the floor, you need to build a launcher capable of doing this; but it need not look nice or have special controls. Once you have built the prototype, you test it to see how well it satisfies the requirements.

Figure 1.6 shows a prototype ready for testing its ring collection, transfer, and launching system. You will use this type of prototype to test the rate of ring collection, its ability to collect rings in awkward positions, ability to transfer rings to the launcher, and ability to launch rings quickly and accurately at targets.

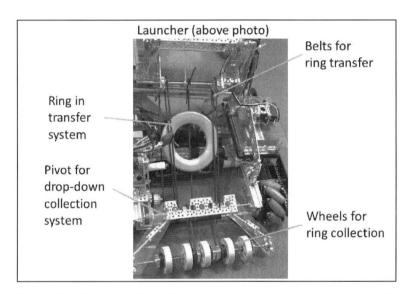

FIGURE 1.6. EXAMPLE RING COLLECTION, TRANSFER, AND LAUNCHING SYSTEM PROTOTYPE

If the prototype robot does not meet requirements, you must determine what to modify to improve performance. This might require minor modifications or a change in the concept for part of your robot. Once the prototype satisfactorily meets requirements, you move to the next stage of design: Solution Completion.

Solution Completion Stage

How do you make your robot ready for competition? You will select off-the-shelf parts, purchase materials, fabricate specialized parts, and assemble the physical robot. You then test the completed robot to ensure that it works as intended.

To build the mechanical structure, you must select structural components and fasteners that support expected loads and allow desired motion of joints. You will buy off-the-shelf parts and fabricate others. Having CAD drawings for the full robot will make the purchase and fabrication of parts much easier, and it will guide your team's assembly efforts. You also need to find space and mount your robot's electronic parts, batteries, and sensors.

Your robot will have electrical devices for powering and controlling its actions. You must select motors and gearing to power movement at desired speeds. Select servos and sensors, install them, and program them to produce precise motion of your robot's mechanisms. You will need autonomous (driverless) and driver-controlled programs to communicate with the robot's sensors, motors, and servos. Batteries and motors must be mounted, and wiring routed to transmit power and signals for operating the robot. You may need shielding to protect electronics from damage and protect operators from hazards. Whew!

Once your robot is assembled, you ought to test it to ensure that it performs as expected. Testing undoubtedly will reveal flaws and cause you to refine the robot and/or its programs. You will need practice matches to test your robot's durability, gather performance data, and give operators practice.

Note that this book presents design principles to guide you in your robot design. To help you understand how these principles are applied, examples are provided, often taken from design activities or design lessons presented to the Robo Raiders FTC 7129 team. Table 1.1 is such an example of summary test data for eight practice matches.

TABLE 1.1. EXAMPLE ROBOT TEST DATA

Scoring Method	Average Time	Scoring Attempts	Scoring Successes	Success Rate
Autonomous: WG in target zone	4.2 s	8	6	75%
Autonomous: Ring in high goal	1.3 s	24	12	50%
Driver-Control: Ring in high goal	5.7 s	72	56	78%
End Game: WG in drop zone	4.4 s	8	7	87%
End Game: Power shot	1.6 s	15	6	40%

Document your completed robot solution with photos, drawings, or sketches that show key features and their functionality. You will probably need several images or views to show different features. You might even remove parts to show some features of the robot. Videos are effective for showing key movements and speed of scoring.

Figure 1.7 shows a robot photo with parts labelled to highlight important features and functions. Note that parts are not simply named, but have explanations built into labels. This detail is important for communicating what your robot can do.

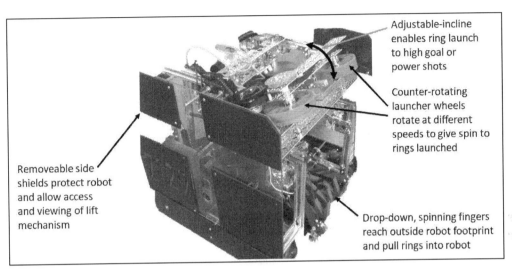

FIGURE 1.7. EXAMPLE PHOTO OF COMPLETED ROBOT SOLUTION

Two video resources and a book chapter prepared for coaching FTC teams provide additional information relevant to this chapter.

VIDEO: Expert Interview: Design Process (6:44 min) https://youtu.be/Uv3xwtos1as
VIDEO: Lecture: Engineering Process Overview (22:53 min) https://youtu.be/1usluFn_T7g
BOOK CHAPTER: Structural Components (26 pp.) *Pre-Engineering Primer, 2nd Edition*, Ch. 4., by Denny Davis

Before moving on to the next chapter, realize that this introduction to robot design is a quick overview of the process. The design process involves much greater detail, using numerous steps to bring more creativity into the design and better judgment to ensure that the design will work.

Figure 1.8 gives you a glimpse of the process steps you will explore in the following four chapters. Your robot design will move from Defining Game Strategy (step 1) down the diagram to Evaluating Completed Robot (step 12). When you see weaknesses along the way or better information becomes available, you might choose to redo earlier steps (design iteration) to improve the design, and then proceed forward. You probably will use iteration many times, each time repeating the same or different design steps.

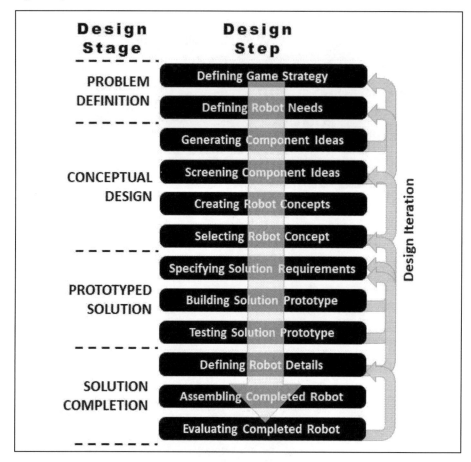

FIGURE 1.8. ROBOT DESIGN PROCESS WITH DESIGN STEPS

You will learn much more about steps in the design process as you study them stage by stage. Chapter 2 addresses the Problem Definition stage, Chapter 3 the Conceptual Design stage, Chapter 4 the Prototyped Solution stage, and Chapter 5 the Solution Completion stage. You will also learn how to document your work in each design step. Review questions (as shown on the next page) are provided to test and deepen your learning about each design step.

Review Questions: Robot Design

1. In your own words, what is the robot design process?
2. What is the purpose of defining needs for the solution?
3. In your own words, what is a conceptual design solution?
4. What is the difference between a solution concept and a completed solution?
5. What is the purpose of design iteration?

ANSWERS TO REVIEW QUESTIONS

1. The robot design process is a systematic process initiated by a game challenge and resulting in an operational robot meeting the team's needs.
2. Defined needs clarify what a robot must be and do to for success.
3. A conceptual design is a solution in its bare bones form, showing key features and functionality that enable it to be successful.
4. The completed solution has all the features of the solution concept plus many details that produce consistent performance, durability, legality in games, and appearance to make it "finished".
5. Design iteration is used to improve the design by revising earlier work to remove weaknesses and produce a design-in-process with greater potential to become an excellent finished product.

Mini-Project: Creating Value from a Waste Material

INSTRUCTIONS: *In small groups, conduct the following activities as an exercise in engineering design. You might choose to share results among groups after each activity or at the end. At the end, discuss how each step contributed to successful design. Then celebrate your design successes.*

PROJECT ASSIGNMENT

Empty boxes and liquid containers can be waste products or resources. Your assignment is to create a useful product out of an empty jug or bottle. See if you can make something that offers value to a recipient (family member, friend, neighbor, etc.) of your product and gives you satisfaction from producing a quality product.

Activity 1: Problem Definition

Think imaginatively about this opportunity to design a new product. Write a statement of the problem as an opportunity, and develop a list of needs that a good solution will address.

Activity 2: Conceptual Design

Brainstorm for ideas for the solution that could satisfy your problem statement and the needs you identified. Then select from among your ideas the one that best meets your stated needs. Describe what this new product will be.

Activity 3: Prototyped Solution

Begin constructing a rough representation of the product you envision. Keep in mind the needs you stated so your conceptual design addresses those needs. Your prototype should have enough features so a potential recipient can give you feedback as to what it yet needs to become a "winner."

Activity 4: Solution Completion

Take time to enhance and complete your product to make it as you would present it to the recipient. Make it look nice and accomplish the stated needs. It should be a product the recipient would value and you will take pride in.

Understanding is a wellspring of life to one who has it.

Chapter 2: Problem Definition (Stage 1)

The engineering design process uses several steps or activities during each design stage to develop high quality intermediate and final design products. Figure 2.1 shows the four design stages—Problem Definition, Conceptual Design, Prototyped Solution, and Solution Completion—each with multiple steps.

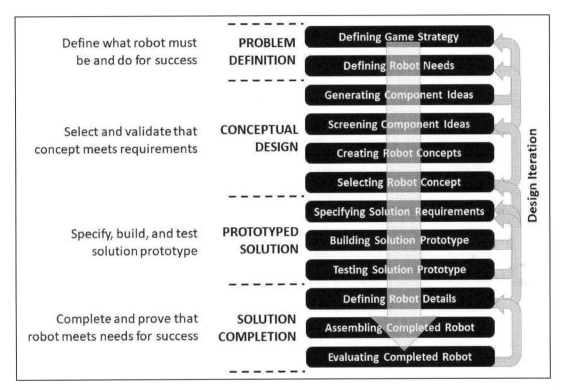

FIGURE 2.1. STEPS IN THE ENGINEERING DESIGN PROCESS

The next four chapters, address the four stages and discuss each of the twelve design steps in detail. You might use these four chapters to guide you through creative and judgmental activities needed to develop a high-quality robot design solution in a period of several months.

Step 1: Defining Game Strategy

Defining Your Game Strategy

A game strategy is a plan for competing in a game. Your game strategy will be your attempt to maximize your game score and represent your team well on a field with competing and cooperating robots. Thus, you must understand how best to score and how to demonstrate your robot's strengths during a match.

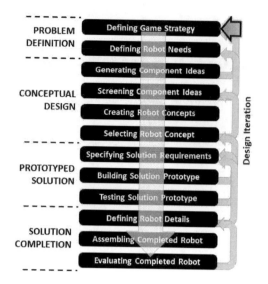

In FIRST® FTC competitions, the game is divided into three time periods: autonomous, driver-controlled, and end game. Your team needs a strategy for each period of the game, considering the scoring options and challenges of that period. You should thoughtfully develop your game strategy based on the likelihood of scoring success and points available for different scoring options.

A systematic way to evaluate scoring options is illustrated in Table 2.1 for a 30-second autonomous game period. For each scoring option, you will estimate the time required to score and the probability of success in one attempt at scoring those points. You will use this information to calculate probable (average) points earned for that scoring method: **Probable Points = Probability × Points**. Scoring efficiency is the number of probable points scored per second, **Efficiency = Probable Points / Time**, a measure of your robot's potential for rapid scoring.

TABLE 2.1. ESTIMATED PERFORMANCE FOR AUTONOMOUS SCORING OPTIONS

Autonomous Scoring Option	Points	Probability	Time	Probable Points	Efficiency
WG delivered to target zone	15	0.70	12 s	10.5	0.9
Robot on launch line after WG in target	5	0.90	4 s	4.5	1.1
Ring #1 launched into high goal	12	0.40	5 s	4.8	1.0
Ring #2 launched into high goal	12	0.35	2 s	4.2	2.1
Ring #3 launched into high goal	12	0.30	2 s	3.6	1.8
Collect and score ring into high goal	12	0.10	10 s	1.2	0.1
Robo on launch line after ring goal	5	0.90	2 s	4.5	2.2

From the data in Table 2.1, you will observe the following:

1. The largest number of probable points in one scoring attempt will occur from delivering the wobble goal (WG) to the target zone. Thus, you should attempt this scoring method if your team has capabilities to design for this method of scoring.

2. Ample time is available in the 30-second autonomous period to score wobble goal, launch 3 rings, and park on launch line.

3. If your team has capabilities to collect and launch additional rings autonomously, that might give you 1 additional point but consume 10 seconds. This attempt would probably not be possible due to time required, if you attempt it after scoring the wobble goal and launching 3 rings that are initially on the robot.

4. Suppose your team is pondering options for the end of the autonomous period: try scoring an autonomously collected ring vs. parking on the launch line. Greater probable points are predicted for parking on the launch line.

Based on this information, your team probably will construct this game strategy for the autonomous period:

1. Deliver the wobble goal to the target zone.
2. Launch 3 rings (initially loaded on the robot) at the high goal.
3. Park on the launch line.

You will conduct a similar analysis of scoring options for the driver-controlled period and the end game period to define strategies for these periods. For each period and overall, consider probable points earned, scoring efficiency, and your team's capabilities to design and operate the robot for the planned scoring attempts. Also think about what other teams on the field might do and how well your drive team can respond when unexpected conditions occur during a match.

You should be ready to alter your game strategy in a match, and most certainly revise it as the season progresses and your drive team and robot's capabilities improve.

Documentation for Defining Game Strategy

When you have completed this step, carefully document your team's game strategy and your rationale for this strategy. An example is shown in Figure 2.2. The horizontal dimension shows the timeline of a match. Rectangles show the durations of scoring attempts during each period of the game. With each type of scoring attempt, the success rate, time expended, and probable points earned are given. This information helps the reader see where points are earned, where time is expended, and the contribution of different scoring methods to the team's overall score. Date your strategy so you can track changes in robot performance if you change your strategy later.

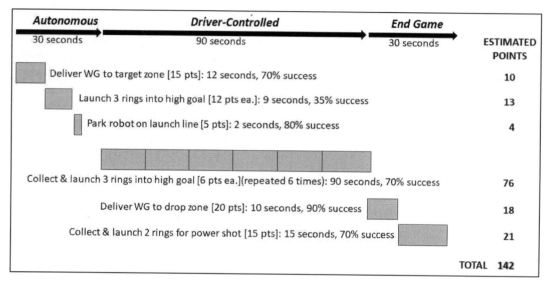

FIGURE 2.2. DOCUMENTATION OF GAME STRATEGY

Resources for Defining Game Strategy

For more information about developing a game strategy, watch the first part of a video developed for coaching FTC teams:

  Strategy + Needs (33:20 min)   https://youtu.be/79-BeQY59ZM

When you are ready to develop a game strategy to test your understanding, pick a robot game of the past and analyze the game. A worksheet is provided on the following page for recording your scoring methods and estimates of scoring successes. Once the sheet is completed, consider the best scoring options and the best order in which to make scoring attempts. You might wish to attempt easy scoring early in a game period so you accumulate points before failures occur that might prevent you from making tougher scores.

GAME STRATEGY WORKSHEET

Prepared by:　　　　　　　　　　　　　　　　**Date:**

Scoring Method	Points	Probability	Time	Probable Points	Scoring Efficiency

Column 1: List specific ways to score being considered, perhaps after a specific action.
Column 2: List the number of **Points** earned from a successful attempt.
Column 3: Estimate the **Probability** that the described scoring attempt is successful.
Column 4: Estimate **Time** required to score, from start to finish of one attempt.
Column 5: Calculate **Probable Points** as the product of **Points** and **Probability**.
Column 6: Calculate **Scoring Efficiency** as ratio of **Probable Points** divided by **Time**.

Review Questions for Defining Game Strategy

1. In your own words, what is a game strategy?
2. What is the purpose of the game strategy?
3. What information is used to define a game strategy?
4. How do probable points help you predict your match score?
5. How does scoring efficiency affect a game strategy?
6. Why might a game strategy change for a different competition?

ANSWERS TO REVIEW QUESTIONS

1. A game strategy is a sequence of robot actions intended to play the game successfully and hopefully achieve the best score possible.
2. The game strategy is defined so that the team agrees what the robot will attempt to do in a game. It is a starting point for designing a robot to carry out these actions.
3. A game strategy is built upon a selection of game scoring attempts, with the probability of success and time to score in each scoring attempt.
4. Probable points are an average of how many points are scored in multiple attempts of one kind. Combining probable points for all methods planned will give you an estimate of total game points.
5. Scoring efficiency shows how quickly points can be earned with a scoring method. Methods with high scoring efficiencies should not be ignored.
6. Estimates of scoring probabilities and time to score will change as a robot and drive team improve. This changes the probable points and scoring efficiencies for methods, which likely will change the team's game strategy.

Step 2: Defining Robot Needs

Defining Your Robot Needs

Define your robot needs from problems or opportunities (challenges) in your game strategy or in other planned uses of your robot (such as demonstrations in a community). Since your game strategy and robot uses may change as the season progresses, your robot needs might best be defined for early, intermediate, and advanced competitions in the season.

To define robot needs, describe functionality and features your robot must have to meet challenges (i.e., to score points). You might define crucial movements, speed, accuracy, or interactions with game elements. Some needs might not be specific now, but they will become clearer as your design progresses and your understanding increases.

Table 2.2 presents a set of needs based on the previously defined autonomous game strategy. Column 1 identifies challenges for scoring using each of the scoring methods selected in the game strategy. Column 2 defines corresponding robot needs derived from the challenge. Note that needs identify what software and hardware of the robot must do to meet the scoring challenge.

TABLE 2.2. EXAMPLE DEFINITION OF ROBOT NEEDS FOR AUTONOMOUS PERIOD

Problem/Opportunity	Statements of Corresponding Robot Needs
Autonomous: Robot must sense number of rings and deliver wobble goal (WG) to corresponding target zone. (15 pts)	• Sensor and software decide if zero, one, or four rings in starter stack. • Pre-programmed software directs robot to drive WG to proper 22.75-inch-square target zone on field. • Robot arm grabs WG, holds WG in transport position during transit, and releases it in standing orientation upon command from software.
Autonomous: Robot must score three pre-loaded rings in high goal. (36 pts total)	• Pre-programmed software directs robot to drive behind launch line, turn to face goals, and launch 3 rings. • Drivetrain moves robot to launching position 65±4 inches from goal and ±2 degrees from goal center. • Launcher hurls rings to hit high goal 65 inches away centered 35.5 inches from floor, 5 inches tall, and 16 inches wide.
Autonomous: Park on launch line. (5 pts)	• Pre-programmed software directs robot to drive onto launch line. • Drivetrain moves center of robot to distance 60±5 inches from goals.

Needs also must be defined for driver-controlled period scoring and for end-game scoring. In addition to these scoring-related needs, your team might add needs that come from practical issues, game rules, or robot image. Examples might include robot balance to give good traction, operator visibility of robot functions, robot weight or height to satisfy rules or enhance maneuverability, placement of and access to components for maintenance, materials available to the team, or skills of the team.

Documentation for Defining Robot Needs

You might choose to document robot needs in two different ways. First, document the challenge-to-needs definition activity as shown in Table 2.2. Then reorganize needs by robot component and present them as shown in Table 2.3. Note in Table 2.3 that multiple needs may relate to one component.

TABLE 2.3. COMPILATION OF ROBOT NEEDS FOR AUTONOMOUS PERIOD

Robot Component	Statements of Corresponding Robot Needs
Drivetrain	• Drive to 22.75-inch-square target zone for WG delivery. • Drive to launching position 65±4 inches from goal and face 0±2 degrees from goal center. • Drive center of robot to position 60±5 inches from goals.
Ring sensor	• Sense starter stack and determine if zero, one, or four rings are present.
Wobble goal arm	• Grab WG, hold WG in transport position during transit, and release it in standing orientation upon command from software.
Pre-programmed software	• Direct robot to field coordinates of target zone matching the number of starter stack rings. • Direct robot to field coordinates and orientation for launching rings. • Direct robot to field coordinates for parking on launch line. • Direct launcher to load from internal stowage and launch three rings in sequence at high goal.

Resources for Defining Robot Needs

Additional discussion of needs definition is found in the second part of the following video developed for FTC teams:

Strategy + Needs (33:20 min)   https://youtu.be/79-BeQY59ZM

A worksheet on the following page is useful for defining robot needs from game challenges.

WORKSHEET FOR NEEDS DEFINITION

Prepared by: Date:

Problem or Opportunity	Specific Robot Needs Definition

A **Problem or Opportunity** is a general statement of desired scoring or other success.
A **Need** is a specific statement that defines essential performance of parts of the robot, automated robot controls, or operator control actions that command robot responses.

Review Questions for Defining Robot Needs

1. In your own words, what is a robot need?
2. What is the purpose of defining robot needs?
3. What information is used to define a robot need?
4. What makes a good statement of a robot need?
5. What might cause a robot need to disappear or change over time?

ANSWERS TO REVIEW QUESTIONS

1. A robot need is an action or response of a robot component essential to achieving a desired success, such as scoring in a match.
2. Robot needs define specific actions or tasks that a component must perform, thereby providing a basis for judging ideas for those components in the design process.
3. A robot need is defined by breaking a success (e.g., scoring success) into factors (actions or responses) that achieve that success. Needs define what robot software, mechanisms, and operators must do for the success.
4. A good needs statement identifies the robot component and what it must do, being as specific as your understanding enables.
5. A robot need might change if the scoring method changes, or the robot components change. If the robot is simplified or a component is removed, the corresponding needs might disappear.

Listen to counsel and receive instruction so you may be wise in your latter days.

Design Review: Problem Definition Stage

Before progressing to the Conceptual Design stage of design, you must be sure that your work to this point is adequate to provide a credible foundation for the design of your robot. You need to be able to confidently say "Yes!" to the following questions:

    Does your team fully understand the game and robot judging criteria?

    Is your strategy ambitious and yet achievable for your team?

    Does your problem statement communicate inspiring robot goals?

    Do your robot needs define clear targets to guide your robot design?

Plan a Problem Definition design review to address these types of questions. Invite one or more people to probe your work to test its credibility. Ideally, these reviewers will be fair in judgments, willing to challenge your work, and knowledgeable about the robot game and competition context. Ask them to be critical and yet helpful in making your design work better.

The following are suggested prompts or questions that can be used by reviewers to probe your Problem Definition work:

- List available scoring methods in the robot game in order of your team's priorities for scoring. Explain why you prioritized each method as you did.
- Explain how judging criteria relate to your robot's design.
- Explain your game strategy in terms of scoring, judging, and capabilities of your team.
- State your robot design challenge in terms of what you expect your robot to be and do.
- List specific robot needs and targets that must be met for your robot to be successful.

Ask your reviewers to rate your Problem Definition and provide suggestions for improvement to your Problem Definition. The worksheet on the following page is useful for gathering this feedback on your Problem Definition.

FEEDBACK FOR PROBLEM DEFINITION DESIGN REVIEW

Review for (team name):

Reviewer names:

Date of review:

Information provided in advance of review:

Please rate the review quality in each area (place X in appropriate column)

Area of Review	Weak	Acceptable	Outstanding	Notes
Understanding of robot game				
Understanding of judging criteria				
Soundness of game strategy				
Boldness in game strategy				
Energizing nature of problem definition				
Breadth of robot needs defined				
Challenge posed by target values				
Overall team readiness to proceed				

Comments on review process:

Suggestions for improving Problem Definition:

In a multitude of counselors there is safety.

Chapter 3: Conceptual Design (Stage 2)

The Conceptual Design stage of the design process seeks to create a solution concept that will meet all needs defined earlier. In this stage, you will explore in depth four design steps:

Generating Component Ideas

Screening Component Ideas

Creating Robot Concepts

Selecting Robot Concept

Once you have completed this stage of design, selected the conceptual design solution, you will be ready to proceed to the Prototyped Solution stage of robot design.

At the end of the current design stage, you will be guided through a Conceptual Design review. When your selection of a conceptual design is judged sound, you are ready to prototype and test your proposed solution.

Step 3: Generating Component Ideas

Generating Your Robot Component Ideas

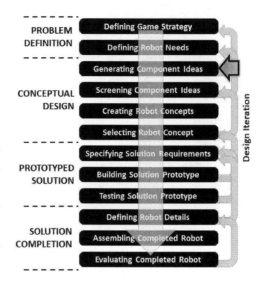

Ideas are generated to identify multiple and possibly creative ways that robot design needs can be met. Because a robot is complex, your team will benefit from decomposing the robot into components, probably according to functions it must perform. Then these components can be discussed as systems within the robot, and in fact, each system can be considered as a small design project. Example systems for the game robot discussed above might be drivetrain, collection system, launcher, and wobble goal manipulator.

For any of the systems that will be part of a design, you must generate ideas to provide options for the design of that system. Ideally, you will identify many ideas for each system—some that are familiar and known to work, others that are untried and have potential to work, and even some that are wild and unlikely to work. A wide range of ideas can spark creativity and reveal possibilities for combining part of one idea with another to formulate novel or improved solutions.

When generating ideas, start by looking at ways your team or others have addressed the same or similar challenges in the past. Look online, search trade magazines, visit trade shows, look at tools and appliances at home, and analyze mechanisms in farm or materials handling machinery. Look seriously at robots from years gone by because they had to work under conditions like your competitions. Don't hesitate to talk to people and ask them for ideas. This is an opportunity to contact engineers and tinkerers nearby, building valuable relationships for future interactions.

Be sure to record every idea you generate and its source so you can go back and dig for deeper understanding. Ask your team members to privately brainstorm and record their ideas for specific systems. Then gather teammates together to share ideas and, hopefully, springboard to new and better ideas. Be careful that you do not criticize one another's ideas too early because that can send a "chill" over the discussion and shut down creativity. Encourage wild ideas that may have merit.

Once you have a bunch of ideas, identify relevant categories and group ideas under these categories. For example, a team's collection system ideas might fall into categories of scoops and wheels. When you discover that you have completely ignored categories of fingers and suction devices, you might begin a new effort seeking ideas for these categories. Note that you also might look for ideas that address more than one of your systems, such as ideas that combine collection and launching.

Documentation for Generating Component Ideas

Documenting idea generation for each system in your robot decomposition might best occur at two levels. Level 1 is capturing all ideas and their sources. This is vital for properly crediting ideas and for capturing all idea for later consideration. Table 3.1 shows a suggested format for documenting this rich resource of ideas.

TABLE 3.1. EXAMPLE DOCUMENTATION FOR IDEAS AND SOURCES

Launcher Idea	Description	Source
Skeet shooter	Cocked arm with horizontal swing	Jose, Joe's neighbor
Crossbow	Cocked string thrower	Sally
Baseball pitcher	Spinning 2-wheel thrower	Xavier
Trebuchet	Spring-loaded with vertical plane swing	

The second level of documentation is a more detailed description of ideas, especially those needing clarification. Your description might include a sketch, cardboard or Lego model, or picture of an object depicting the idea—a representation that enables the team to understand what it means. The process of describing a given idea might spark new ideas or might just reveal that variations of this idea should be considered as well. Figure 3.1 illustrates how an idea for a collection system might be documented.

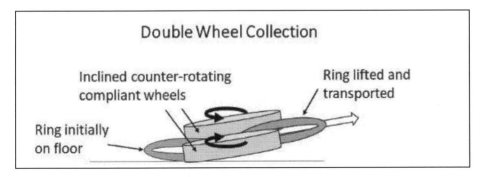

FIGURE 3.1. EXAMPLE DOCUMENTATION OF COLLECTION SYSTEM IDEA

Resources for Generating Component Ideas

Additional discussion of idea generation is found in the first part of a video created for FTC teams:

Generating + Screening Ideas (31:55 min)   https://youtu.be/7pecwYE3MMI

Review Questions for Generating Component Ideas

1. What is the purpose of the Generating Component Ideas step?
2. What types of ideas are desired from idea generation?
3. How can other teams' robot game videos be useful for generating ideas?
4. What are other good sources of ideas for mechanical systems?
5. What guidelines are important for brainstorming?
6. What information should be recorded for ideas generated?

ANSWERS TO REVIEW QUESTIONS

1. Ideas are generated to provide options for different parts of the robot. Novel ideas can add creativity and generate excitement about the design.
2. A range of ideas is desired. You want practical ideas and some that are wild. You want lots of ideas and widely varied ideas, so they bring opportunities for innovation.
3. Videos of other robots expose you to others' creativity, show you what has worked in similar situations, and spur you on to be better than the competition.
4. Other good sources of ideas are equipment for doing similar tasks, creative people, patents, and websites or books that show mechanisms for doing different things.
5. Guidelines for brainstorming: invite creativity and lots of ideas; ask for ideas from everyone, even shy people; capture every idea; avoid criticism (verbal or body language) that could cause people to hold back ideas; incentivize the largest number and most creative ideas.
6. Record a name for the idea, brief description if needed, and source of the idea.

Step 4: Screening Component Ideas

Screening Your Robot Component Ideas

Your idea generation should have yielded many ideas for each system in your robot. Now, to focus your design efforts, pick from this list those ideas that have the greatest potential to meet your needs. At this early point in the design process, as you set aside ideas that are not a good fit to needs, be careful not to discard all ideas with potential to bring creativity to your design. To make idea screening effective, establish a set of criteria that you can use to judge the merits of ideas.

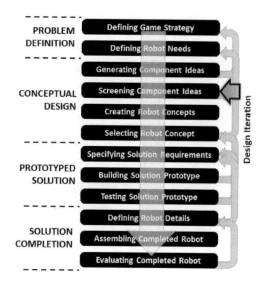

What might be suitable criteria for keeping an idea? Consider the following categories of criteria:

- Creative: has potential to lead to valuable innovation
- Functional: performs the function required of the system
- Compatible: will interface well with other systems in the robot
- Plausible: can be designed, built, and used effectively by the team
- Durable: can be made to work consistently without failure

For each system, define a set of criteria befitting that system for use in the idea screening process. For example, the **functional** criteria for a collection system might be (1) able to collect quickly, (2) able to reach into corners, (3) will not jam, and (4) able to collect one from among many. For your team, **plausible** criteria might be (1) able to build with available tools, (2) able to control with our team's programming capabilities, and (3) will not cost more than $50.

If you have lots of ideas, you might want to first screen with 3 to 5 general criteria such as the five listed above, then afterward screen the ideas kept using more specific criteria. Remember that the purpose of screening is to find the best ideas for each system so they can be considered further, yet to keep ideas with potential for innovation.

The screening process needs to be done systematically so that ideas are considered fairly, and personal prejudices do not prevail. Use a screening matrix as shown in Table

3.2 to facilitate this process. Ideas are listed in column 1 by a name or a description that differentiates each idea. Place criteria for screening as headings above other columns. Screening is done by rating each idea (possibly rate using 0 = poor, 1 = fair, 2 = good, or 3 = excellent) in each column to indicate how well it fits the respective criterion. Row totals provide a rating total for the ideas, with higher being better.

TABLE 3.2. IDEA SCREENING EXAMPLE, SIMPLE SCREENING

Ideas for Ring Collection System	Criterion 1: Creative	Criterion 2: Functional	Criterion 3: Compatible	Criterion 4: Plausible	Criterion 5: Durable	TOTAL
Scoop 1, no active elements	0	1	1	3	3	8
Scoop 2, horizontal-angled brush above	2	3	3	2	2	12
Scoop 3, dual vertical-axis brushes	1	3	3	2	2	11
Claw 1, operator-guided	2	1	2	1	1	7
Claw 2, sensor-guided	3	2	2	0	1	8

Assign ratings for an idea after a discussion among involved team members, especially those knowledgeable about the ideas and robot needs. Use these discussions to help members understand ideas; listen to everyone's voice so decisions consider all perspectives and are owned by your team.

From the screening decision matrix above, you can see the following:

- Scoop 1 is not creative, may not be effective in gathering rings, and will not move rings toward the launching system; it is not a strong idea.

- Claw ideas afford creativity but are not durable or feasible with team skills and may not be effective at collecting rings; these are not strong ideas.

- The most promising ideas are scoops 2 and 3 with powered brushes to pull rings into and push them out of a scoop. These ideas might offer some opportunities for creativity and appear to be functional and plausible.

Your design team should screen ideas for each robot system. As mentioned above, you might need to do more than one screening if you have many feasible ideas; if so, use more specific criteria for second screenings. For each system and screening, create a table that shows how ideas are scored.

Documentation for Screening Component Ideas

Documenting the idea screening process might use tables of two types. The first is the screening matrix, as shown above, for each system. In addition, you might create a summary such as Table 3.3, which presents the ideas retained for all systems, capturing all ideas kept for further consideration.

TABLE 3.3. SUMMARY OF IDEAS KEPT AFTER SCREENING FOR ROBOT SYSTEMS

System	Ideas Retained
Drivetrain	Holonomic chassis with Mecanum wheels at 45 degrees from forward
Ring collection	Scoop with wheels positioned at both sides and inclined slightly
	Scoop with brush above scoop to pull in and push out
Launcher	Paired counter-rotating compliant wheels that throw rings
	Single rotating wheel that spins and launches rings
	Spring-loaded arm to throw game rings
Wobble goal lift	Arm that pivots over-the-top from one pivot point
	4-bar mechanism that moves over-the-top of the robot
Wobble goal grabber	Scissor with rubber fingers that pinch the wobble goal stem
	Swinging arm that reaches beyond, pulls in, and hugs wobble goal

Once you have completed idea screening for each of the robot systems, you are ready to progress to the next step in the design process. However, keep in reserve any good ideas that you screened out, because you might find later that you need more good ideas after your "best" ideas failed.

Resources for Screening Component Ideas

Additional discussion of idea screening is found in the second part of a video created for FTC teams on idea generation and screening:

Generating + Screening Ideas (31:55 min)   https://youtu.be/7pecwYE3MMI

Review Questions for Screening Component Ideas

1. Why should generated ideas be screened?
2. Who should be involved in screening ideas?
3. What might be useful criteria for screening robot drivetrain ideas?
4. Why should ideas that are creative be retained at this point if they have some possibility of meeting needs?
5. What information should be documented for an idea that is retained after screening?

ANSWERS TO REVIEW QUESTIONS

1. Ideas should be screened to eliminate those that do not have potential to meet needs of robot systems. Keeping too many ideas for active development makes the workload excessive and the design process bogs down.
2. Multiple people should do the screening to reduce personal bias. People closest to the idea and closest to the system for which it is identified should be involved because they have a stake in the system's success. If the whole team is involved, this might slow the process but would also increase possibilities of catching weaknesses and would gain whole team buy-in.
3. Screening criteria for drivetrain ideas might include durability, mobility, number of motors required, ease of programming, traction, and innovation.
4. Creative ideas are valuable to the team because they set your robot apart from others and may qualify you for an innovation award. Even if a creative idea has some weaknesses, it may be refined later to make it feasible and innovative.
5. For each idea kept after screening, document the idea to make it understandable to others. This might simply require a name that people understand, but it instead might require a description of key parts and how they work, or even a sketch.

Step 5: Creating Robot Concepts

Creating Robot Concepts from Ideas

The next design step is creating concepts for the entire robot, a vital part of the Conceptual Design stage of design. Conceptual design, also called systems design, is the design stage where the major systems are selected for incorporation into the robot and further development. Your goal in conceptual design is finding systems that work well together and might even produce synergies when they are combined.

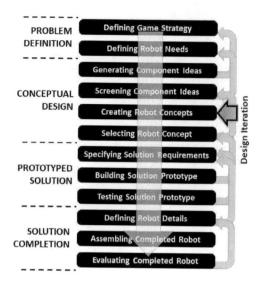

Conceptual design can be the greatest determining factor in robot design success. A poor selection of systems can lock your team into a path that will prevent your robot from becoming excellent. If you select a poor robot concept, you will invest much energy in something that does not work or is boring, leading to weak robot performance and possibly demoralizing your team.

To define a robot concept, you will choose one idea for each of the robot's systems, which when brought together represent a potential robot. For example, a robot concept might be the combination of a drivetrain idea, a collection system idea, a game element delivery system idea, an autonomous scoring manipulator idea, and an end game scoring manipulator idea. You should be able to compose several feasible concepts from the ideas you kept after idea screening done earlier.

Because concepts require a set of systems, a table or chart is useful for generating and recording concepts so that all systems are included in each robot concept. Engineers often use a morphological chart or "morph chart" to catalog concepts, one at a time, ensuring that all systems are identified in a concept.

Table 3.4 is a partial morph chart for a robot with five systems, showing ideas that were retained for systems after screening. Note that the number of possible concepts is the product of the numbers of ideas retained for each system. When you have a large number of possible concepts to consider, think critically about how many concepts your team can consider reasonably so you do not keep too many ideas.

TABLE 3.4 PARTIAL MORPH CHART FOR CONCEPTS FOR 5-SYSTEM ROBOT

Concept	Drivetrain	Collection	Launcher	Lift	Grabber
1	Mecanum	Side wheels	2 wheels	Swinging arm	Hinged fingers
2	Mecanum	Brush above	2 wheels	Swinging arm	Hinged fingers
3	Mecanum	Side wheels	1 wheel	Swinging arm	Hinged fingers
4	Mecanum	Brush above	1 wheel	Swinging arm	Hinged fingers
5	Mecanum	Side wheels	2 wheels	4-bar	Hinged fingers
6	Mecanum	Brush above	2 wheels	4-bar	Hinged fingers

Documentation for Creating Robot Concepts

A completed morph chart of the type shown in Table 3.4 is a suitable method of documenting concepts created. To make the ideas (words) understandable, you might want to reference definitions of ideas you record elsewhere or give descriptions along with the morph chart. Figure 3.2 provides more visual way to show the ideas retained for each function, combinations of which are the robot concepts presented in Table 3.4.

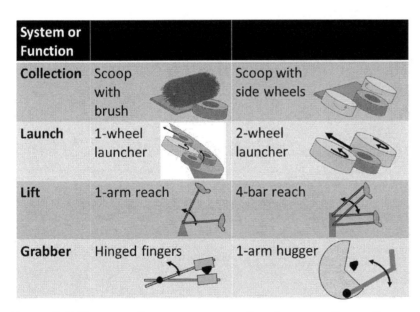

FIGURE 3.2. PICTORIAL REPRESENTATION OF KEPT IDEAS FOR ROBOT SYSTEMS

Resources for Creating Robot Concepts

For further discussion of creating concepts, see the first part of the following video:

Creating + Selecting Concepts (39:18 min)   https://youtu.be/BwdZP27fmt0

Review Questions for Creating Robot Concepts

1. What is a design concept?
2. How does a concept differ from a completed design?
3. Give an example for a robot conceptual design.
4. What should be documented for each robot design concept that is created?

ANSWERS TO REVIEW QUESTIONS

1. A design concept is a general description of a robot that includes designation of major systems in the robot. It identifies the types of systems useful for performing the critical functions for success, but not specific or detailed parts.
2. A completed design goes beyond the concept to defining every part, connection, dimension, and software feature to control the robot. The completed design should be detailed enough for a person to build the robot ready for operation.
3. A robot concept will include a definition of the major systems in the robot. An example is a robot with tank drive, a street sweeper collection system, a slider with bucket dump for delivery of game elements, and a rotary set of fingers to spin an end game crank.
4. Documentation of a concept should include drawings or sketches of the conceptual design to show what systems are included and how they are configured to fit in the robot. These should be complemented by a naming or labelling of the components and description of how they work to meet the robot needs.

Step 6: Selecting Robot Concept

Selecting Your Best Robot Concept

After having defined concepts for the robot, your design team can proceed to select the best concept for further development. You will rate each concept against criteria specific to the overall robot design. Before you begin rating, your team must carefully identify a set of criteria and the importance of each to the design. Some example robot criteria and their justification are summarized in Table 3.5.

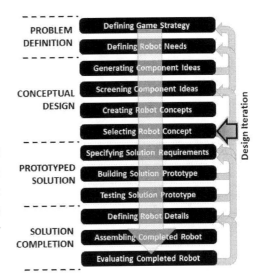

TABLE 3.5. ROBOT DESIGN CRITERIA AND THEIR JUSTIFICATION

Criterion	Justification
Innovation	Distinguishes your robot, increases design challenges and team learning
Speed	Increases scoring in available time, draws attention to your robot
Accuracy	Increases scoring probability, gives evidence of good design and good control
Durability	Keeps robot operating, reduces repairs, shows good design
Simplicity	Shows good design, easier to operate, reduces many problems
Repairability	Enables repair between matches, so no matches are played in disrepair
Manufacturability	Saves time and cost in building, enables high quality assembly
Cost	Staying within budget, allowing team to invest in other efforts

An important part of rating a concept is determining how well the individual systems and their combination meet the needs defined earlier. Rating concepts might be a simple act of judgment based on previous experience, but it might instead require a type of analysis that predicts some aspect of performance. Table 3.6 shows examples of performances and appropriate methods of analysis.

TABLE 3.6. EXAMPLE METHODS OF ANALYSIS FOR JUDGING PERFORMANCES

Performance Type	Method of Analysis
Motor torque, speed	Motor curves showing torque vs. speed matched to load conditions
Travel speed of wheel	Calculate: travel speed = π x diameter x rotational speed
Traction, friction forces	Measure or look up friction coefficient; friction = coefficient x normal force
Load torque	Calculate: torque = force x lever arm
Trajectory of projectile	Calculate from initial velocity and incline angle; use physics equations

Movement of linkages	CAD drawing or physical model
Object collection	Physical model or storyboard of sequential positions
Object transport	Physical model or storyboard of sequential positions

Use a design decision matrix or Pugh matrix to systematically compare and document design decisions for robot concepts. In the Pugh matrix, each concept is rated against criteria weighted by their importance to the design. Because rating many concepts is time-consuming, you might rate them in two passes: first a quick rating using the most important criteria, then detailed rating of the best concepts using all relevant criteria.

To increase separation among concepts, rate them using 1-3-9 values for fit to each criterion, with 1 being low, 3 being good, 9 being excellent. In earlier screening of ideas, you focused on how well ideas fit needs for a system; here concept rating focuses on how well systems in the concept fit and work together to make a successful robot. Table 3.7 is an example Pugh matrix rating five potential robot concepts.

TABLE 3.7. PUGH MATRIX FOR SCORING ROBOT CONCEPTS

Criteria	Importance	Side-wheel collection, 2-wheel launcher, 4-bar with	Top brush collection, 2-wheel launcher, 4-bar with	Top brush collection, 2-wheel launcher, 4-bar with	Side-wheel collection, 2-wheel launcher, 4-bar with	Same as previous + screw transfer to launcher
Innovation	3	3	3	3	3	9
Speed	2	1	1	3	3	3
Accuracy	3	1	1	3	3	9
Durability	2	1	1	3	9	3
Simplicity	2	1	1	3	9	3
Repairability	2	3	3	3	3	3
Manufacturability	2	3	3	3	3	3
Cost	1	3	3	3	3	3
PRODUCT SUM		33	33	51	75	87

In this matrix, you find that the fifth concept scored highest, indicating that it is the best concept. Note its high score (87) is the result of excellent ratings for two particularly important criteria: innovation and accuracy. The next best concept scored 75. Its strong ratings were for durability and simplicity. Because these two top concepts differ in only one feature, your team might decide to continue developing both concepts until the performances of both are evaluated more fully. Keeping these two concepts offers the potential for advancing an innovation if it proves to perform well, and it should cost little for the added work.

Documentation for Selecting Robot Concept

Documentation of concept selection should show the process and rationale for making the selection. You might use the Pugh matrix to show your process and outcomes of the concept rating. Full documentation requires that you explain the rating method and define the rating scale you chose to use. Explain how you interpreted the results and decided which concept(s) will be carried forward, along with any implications this choice will have on design steps ahead.

As you prepare this documentation, you can be rethinking your reasoning for the robot concept; it can also grow member buy-in to the concept decision. Your documentation also will be a valuable resource for a design concept review because the reviewer sees your team's reasoning and your final decision.

Documentation of your design concept surely will include visual representations of the concept. You might use sketches, CAD (computer-aided drawings) of the concept, or photos of a mock-up used in its evaluation. Be sure to label and explain important features of the concept and its movements.

Selection of the best concept completes the Conceptual Design stage of design. Your team should now have a robot skeleton upon which to build the flesh and intelligence to make it into an awe-inspiring robot.

Resources for Selecting Robot Concept

Additional discussion of concept selection is found in the last part of the video prepared for FTC teams:

Creating + Selecting Concepts (39:18 min)   https://youtu.be/BwdZP27fmt0

Another resource that discusses prototyping and can help with evaluating concepts is the second part of the following video:

Requirements + Prototyping (31:19 min)   https://youtu.be/069r6gmWBgg

Review Questions for Selecting Robot Concept

1. Why is selecting the best concept important at this point in the design process?

2. What selection criteria are most important for a robot striving to gain recognition by its game performance?

3. What criteria are most important if the team hopes for a judged award based on design quality and innovation?

4. When should more than one conceptual design be selected for development?

5. What should be documented about the concept selection process?

6. What should be documented about the selected concept?

ANSWERS TO REVIEW QUESTIONS

1. Selecting the best concept enables your team to focus design effort on the selected concept. Making this selection now saves time in gaining focus early so design effort can be more productive working on what will be used.

2. Game performance will be affected most by the speed and accuracy in robot movement and scoring. Durability and repairability will be important to keep the robot performing.

3. Judged awards will be based on innovation and elegance in design. This means simplicity and form fitting function, and effective control (sturdy mechanisms with good programming) will be important. Appearance may be a factor as well.

4. More than one concept might be justified if the team has resources to develop two concepts and the extra work offers high potential for innovation or major improvements. Ideally, a second concept should be continued for a limited period during which one concept is proven superior and it becomes the only one continued.

5. Documentation for the concept selection process should show the criteria used in selection, their weighed importance, and how each concept meets each criterion. This requires definitions of weights, rating options, and how ratings are determined.

6. Documentation of the selected concept must describe the systems in the concept, how they fit together into the robot, and how they produce the performances required for success. Photos or models are necessary parts of the documentation. Essential parts must be identified and described well enough for others to be convinced that this concept will work as intended.

How much better to get wisdom than gold.

Design Review: Conceptual Design Stage

Before progressing to the Prototyped Solution stage of design, you must be sure that your work to this point is adequate to provide a credible foundation for specifying and building a workable prototype solution. To progress, you must show credible evidence that you have done the following:

- Generated creative and feasible ideas for every major system of the robot
- Identified the best ideas for further consideration
- Grouped ideas into workable concepts for the robot
- Selected the best concept for prototyping and finishing as your robot

You should plan a Conceptual Design review to address these questions. Invite one or more individuals to probe your work to test its credibility. Choose people who are unbiased, willing to challenge your work, and knowledgeable about the robot game and competition context. Ask them to be critical and yet helpful in making your conceptual design better.

The following are suggested prompts to be used by reviewers to probe your Conceptual Design work:

- List sources of ideas you used to generate ideas for major systems of your robot. How many came from each source? Which of these ideas bring feasible creativity?
- What are the best three ideas you have for each of the robot systems? Do any of these excite your team with their possibilities?
- What are your most innovative robot concepts?
- Which concepts seem to offer the best scoring potential?
- Which is your best concept overall? What excites you about it? What frightens you about it?

Ask your reviewers to rate your Conceptual Design and provide suggestions for improving it. The worksheet on the following page is useful for gathering this feedback.

FEEDBACK FOR CONCEPTUAL DESIGN REVIEW

Review for (team name):

Reviewer names:

Date of review:

Information provided in advance of review:

Please rate the review quality in each area (place X in appropriate column)

Area of Review	Weak	Acceptable	Outstanding	Notes
Sources of ideas for systems				
Creativity in ideas generated				
Soundness ideas generated				
Quality in ideas retained after screening				
Innovation in robot concepts considered				
Scoring potential in concepts considered				
Potential of robot concept selected				
Understanding of concept selected				

Comments on review process:

Suggestions for improving Conceptual Design:

What is lacking cannot be numbered.

Chapter 4: Prototyped Solution (Stage 3)

Now that you have selected a robot concept, why should you bother with prototyping the robot? Why not just start building the competition robot?

As you will see, building a competition robot demands a huge amount of work, and you have not yet proven that your selected concept will perform as expected. Thus, a wise design team will check that the concept is viable before investing exhaustive effort in building on a possibly flawed concept.

This third stage of design focuses on what is sometimes called "verification" of the design concept: testing (verifying) to be sure that it fulfills all known requirements for success. You will first define those crucial requirements, then prototype and test it to be sure the solution concept merits further development.

You will prototype, not fully build, the robot to save time and cost. Remember, a prototype focuses on only the critical components that show what needs to be evaluated. If you find flaws now, you can correct them before full development of a robot destined for less-than-desired success.

Step 7: Specifying Solution Requirements

Specifying Your Solution Requirements

The first step in the Prototyped Solution stage of design is specifying requirements for the prototyped solution. You need to be thinking, "What must be tested to ensure that this solution concept will do what we want? What are the crucial features and performances that will make our robot successful using our selected conceptual design?"

Your thoughts about requirements probably will go quickly to systems that are part of your conceptual design: drivetrain, collection, manipulators, etc. Each system must perform in certain ways to make your robot successful. Now that you have defined the systems and their interactions in your robot, you can establish requirements for them. Collectively, these solution requirements provide criteria to guide your prototype building and testing.

For example, if a 2-wheel launcher is part of your robot concept, you should specify the speed of the launcher wheels and launcher elevation angle range needed to hit high goal and power shot targets that are at different heights. You might also specify acceptable time delay between launching successive rings. If you are concerned about putting "spin" on launched rings to maintain their orientation while in flight, you should specify rotational speed differences between the two launcher wheels. If you intend to automate robot positioning before launching rings, you should define specifications for accuracy in automated robot positioning.

Specifying requirements often calls for quantitative information—dimensions, speeds, angles, strength, weight, power, etc. To determine a quantitative specification for the robot, your design team might need to create drawings, gather data, or perform calculations.

Let us look at some types of requirements and see how they might be determined. Table 4.1 presents several performances that might be of concern, how suitable requirements might be determined, and example requirements for each. Note that many requirements call for simple calculations based on speeds, geometry, and possibly trigonometry. Some requirements can be determined by conducting a measurement of weight or force or dimension. Others might require building a simple mechanism or conducting a graphical or mathematical simulation of movement.

TABLE 4.1. EXAMPLES OF ROBOT DESIGN SPECIFICATIONS AND THEIR METHODS OF DETERMINATION

Performance	Method of Determination	Requirement
Robot travel speed	Propose that robot travel across 12 ft field in 4 seconds	Maximum robot speed 48 in/s
Lift capacity	Weight of robot estimated to be 40 lb; use factor of safety 2x	Cable lift capacity 80 lb
Lift speed	Want to lift robot 4 inches in 2 seconds	Cable lift speed 2.0 in/s
Launcher speed	Calculation: Velocity for projectile to travel 80 in. with 45-degree launch incline angle	Launcher wheel peripheral speed 240 in/s at 45-degree incline
Ring scoring speed	From game strategy: want to score 15 rings in 90 seconds	Collection, positioning, and launching time: 6 seconds/ring
Collection wheel speed	Propose collected ring travel velocity be 60 in./s from floor into the robot	Wheel peripheral speed 60 in/s
Reach to goal	Arm 6 inches from floor, must reach 24 inches from floor and 6 inches from robot	Arm reach 18 inches up and 6 inches out
Slider pull force	Push 2 lb weight up 60-degree incline and overcome friction coefficient of 0.4	Force = 2 Sin(60) + 0.4 Cos(60) = 1.9 lb → specify 3 lb as load

To prevent errors in requirements, you should have specifications checked by others. You might ask qualified teammates, coaches, or outside experts to review them. It is helpful to talk your way through your reasoning processes and calculations; this helps you find your own mistakes and helps others catch them too. A careless error here can cause you to misjudge verification of your design.

Documentation for Specifying Solution Requirements

Document your requirements specification effort with a list of all your requirements and the basis for each. For each requirement, include both description and quantified target if appropriate. Since many requirements will be used by team members working on different parts of the robot, group appropriate requirements according to the part of the robot to which they apply.

Table 4.2 gives examples of specifications documented for a tank drivetrain that is used on a 12-foot square field with a 45-degree ramp to climb. Each requirement contains a descriptor (column 1) and target value with units (column 2). Some define essential movement magnitudes, and some specify the accuracy of such moves. A traction force is specified based on slope and friction conditions. One requirement specifies a time limit for repair and replacement. When grouping the requirements, some of these will guide the design of mechanical systems, and some will guide the programming of controls to achieve desired responses.

TABLE 4.2. DOCUMENTATION FOR TANK DRIVETRAIN REQUIREMENTS

Requirement	Value	Justification
Speed forward	36 in./s	Cross 12 ft field in 4 seconds
Accuracy forward	2 in.	Within 2 inches of targeted position
Turning speed	180 deg/s	Turn 360 degrees in 2 seconds
Turning accuracy	5 degrees	Within 5 degrees of targeted angle
Control settle time	1 s	After reaching target, settles within 1 second
Speed strafing	12 in/s	Move sideways 3 ft in 1 second
Traction forward	45 lb	Lift 40-pound robot on 45-degree incline with 0.6 friction coef.
Motor replacement	10 minutes	Time between matches might be 20 minutes

Once the set of requirements is prepared, checked, and documented, you are ready to prototype your solution.

Resources for Specifying Solution Requirements

Additional discussion of specifying requirements is found in the first part of the following video prepared for FTC teams:

Requirements + Prototyping (31:19 min)   https://youtu.be/069r6gmWBgg

Review Questions for Specifying Solution Requirements

1. Why is specifying requirements important at this point in designing a successful robot?
2. If robot scoring requires a large reach, what requirements need to be specified to enable scoring?
3. If the team wants the robot to be able to climb a ramp for scoring, what requirements should be specified?
4. What control system requirements should be specified?
5. What should be documented about the requirements defined?

ANSWERS TO REVIEW QUESTIONS

1. Good requirements are necessary to guide next steps in design, which include buying and making parts that cost time and money. The conceptual design identifies types of systems and their integration, which should enable specifications to be defined. Requirements are necessary to define what is essential for movement, strength, access for repairs, and appearance so that design is done right the first time.
2. Requirements for a robot's scoring reach might include reach height required, lateral reach required, accuracy for the reach, and time to complete the reach.
3. Requirements for a robot to climb a ramp might include slope of ramp, height of ramp, climbing force, time to complete climb, and accuracy of end point in climb.
4. Desired performance of the robot's control system must be specified as solution requirements. Examples might be accuracy of driving under autonomous control, stability in positioning of an attachment during scoring, and time to capture an image and use image details to make a decision.
5. Documentation of requirements should be a compilation of requirements, quantitative targets, and bases for the requirements. This information is necessary for others to check your work and be able to identify errors before the requirements are used in the next steps of design.

Step 8: Building Solution Prototype

Building Your Solution Prototype

Why prototype now? Was not prototyping done earlier when ideas or concepts were being evaluated? Yes, you probably did simple prototyping when ideas were being evaluated. At that time, you prototyped to see if ideas were capable of meeting robot needs. Now you will do serious prototyping from which you test your solution's capabilities to satisfy specific requirements placed on different parts of the robot. You are placing high expectations on your selected solution concept.

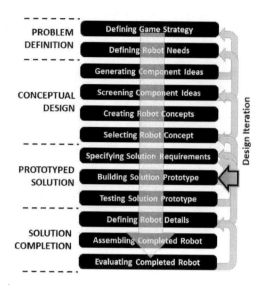

Understand that a prototype is a model or simple representation of something more complex. It is used to answer clearly defined design questions in as simple way as possible. At this point, you will prototype the robot solution with all its systems to determine if it will meet your stated solution requirements. You must know that success is possible before making huge investments in building the competition robot.

A first step in prototyping is identifying what needs to be tested so you know what to include in the prototype. Table 4.3 shows how you might do this. Column 1 lists the systems of a robot. Column 2 summarizes the types of specifications to be tested for the corresponding system. Column 3 identifies features that must be contained in your prototype to test the requirements.

TABLE 4.3. FEATURES NEEDED FOR FULL ROBOT PROTOTYPING

Robot System	Details to be Tested	Features
Drivetrain	Speed of travel across field, accuracy in autonomous travel to chosen position, speed and accuracy of robot turns	Powertrain, wheels, operator controls, autonomous controls, control system algorithms
Ring collection	Success rate and speed of collection, success in retrieving rings from corner	Collection scoop, wheels, and speed control, driver controls
Ring transfer	Success rate and speed of transfer into launching queue	Transfer mechanism with speed control and both driver and autonomous control
Ring launching	Success and rate of launching, success in hitting different targets	Ring launcher with entry as intended, speed control, and any elevation adjustments planned

Wobble goal grabbing	Success rate and time in autonomous grab; success in grab during end game, success in keeping WG during travel, success in releasing WG in vertical position	Grabber attached to drivetrain with full motion and controls as expected for autonomous and driver control
Wobble goal lifting	Success and time to lift wobble goal over fence	Lifting mechanism attached to drivetrain, with driver controls

A major challenge you will face in prototyping the robot is determining how to construct each of the robot systems, where to place them, how to attach them to one another, and how to control them. You need to figure this out to prototype them.

You can often build systems from standard parts available from vendors. For high school robots and similar projects, kits of parts are available as shown in Figure 4.1. Here you find plates, bars, channels, wheels, gears, motors, shafts, hubs, and various connecting pieces. Pieces have pre-drilled holes in patterns that make connections easy. Pieces from such kits also can be purchased separately to meet your needs.

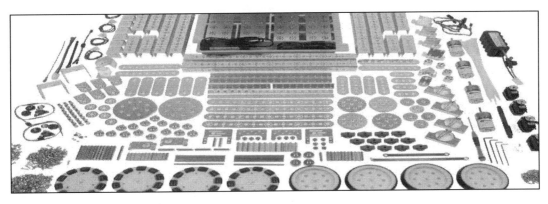

FIGURE 4.1 SAMPLE TETRIX PARTS KIT FOR BUILDING ROBOT PROTOTYPES

Figure 4.2 shows a variety of robot prototypes built from kit parts. Note that chassis, lift, collection, and reach systems can be built with these parts.

Figure 4.2. Simple Structures Created Using Kit Parts [Source: Inner Spark Robotics]

Vendors often make electronic image files of their parts available for use in CAD programs, so you can use these files to construct CAD assemblies. CAD representations of your robot give you an ability to visualize spacing, movement, and points of connection. With these details visualized, you can discover problems that you fix before you waste time trying to build a flawed prototype.

In CAD you can connect parts with desired degrees of freedom (such as rotation about an axis). The CAD allows you to move parts to be sure they don't clash, before building the prototype.

Figure 4.3 shows a CAD image used to represent a robot prototype. Note that it uses some standard parts such as channels and bars with predrilled holes. It also uses original objects created in CAD to represent sliders, spinning brushes, and drive belts. From this CAD, you can see physical arrangements of parts, motion of sliders, and rotation of arms. You can also find places to position electronics, sensors, servos, and motors on your prototype.

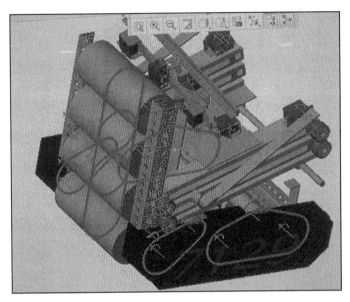

FIGURE 4.3. CAD REPRESENTATION OF A ROBOT PROTOTYPE WITH VERTICAL ROLLER COLLECTOR

Figure 4.4 shows a CAD representation of a different robot prototype. This robot uses vertical extruded rails in its structure, allowing easy sliding adjustments that enable testing of parts in continuous variable positions. This prototype also has two large vertical screws for elevating game elements and a drop-down collection system made of two sets of wheels. CAD enables you to check spacing before making the screws.

FIGURE 4.4. CAD REPRESENTATION OF A ROBOT PROTOTYPE USING SOME EXTRUDED PARTS

A CAD model enables you to see physical sizes of components and identify connection points that allow desired positioning and movement. In CAD drawings you can easily change sizes and shapes of pieces, move attachment points, and check the movement of parts and interferences. You do this modeling before building your physical prototype so you avoid purchasing or manufacturing costly components that may not be needed.

After you know what your prototype requires and you have created a CAD or sketch of it, you need to gather, purchase, or make each part required and begin assembly. Begin by having different members of your team build prototypes of different systems so everyone is involved and systems can be built concurrently. Then assemble the systems and make adjustments necessary for them to work well enough to gather necessary test data.

Minimize time you spend in prototype fabrication but be sure to get the features and functionality that must be tested. Durability, refinement, and appearance are seldom important in a prototype. Prototype parts need not be fancy or light weight, but make them functional and safe.

Documentation for Building Solution Prototype

You should document the process you used to create your robot prototype and document the actual prototype you built. Prepare a table such as Table 4.2 to summarize robot features needed in the prototype. Create another table such as Table 4.3 to record how you determined the features needed in your prototype. Use your CAD drawings or sketches to show how you determined the arrangement and attachment methods for components in your prototype to obtain the features and functions to be tested. You might also explain how you chose to power components for prototype testing. Powering devices do not need to be durable and might be *ad hoc* (such as a variable-speed drill), but they must meet testing needs.

Document your actual prototype with photos. Take photos to show the overall prototype and its key features. Be sure to date and label photos and drawings because they likely will change in the future. Figure 4.5 provides an example of a prototype photo with key parts labelled. Additional close-up photos might be needed to show features or functionality of components.

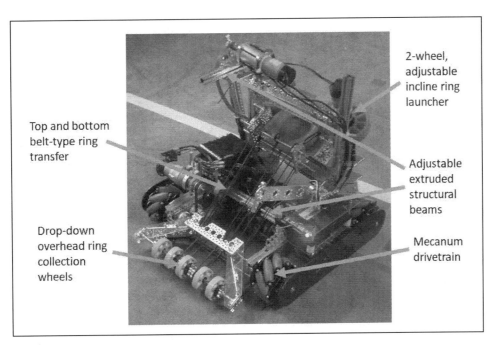

Figure 4.5. Prototype Photo with Parts Labelled

Resources for Building Solution Prototype

Additional discussion of prototyping is found in the last part of the following video prepared for FTC teams:

  Requirements + Prototyping (31:19 min)   https://youtu.be/069r6gmWBgg

Discussion of basic structural components, fasteners, mechanisms, wheels, chains and sprockets, and other power transmission devices can be found in: Chapters 4 through 6 of *Pre-Engineering Primer, 2nd Edition*, by this author.

Review Questions for Building Solution Prototype

1. Why is prototyping the robot solution important at this point in designing a successful robot?

2. What features should be included in the prototype of the robot solution?

3. How is CAD helpful for prototyping a robot?

4. If you have others review your prototype, what should you ask them to do?

5. What should be documented about your robot solution prototyping?

ANSWERS TO REVIEW QUESTIONS

1. Prototyping provides a way to test that the robot will work before you meticulously build the finished robot. You should now find flaws before wasting valuable resources on a robot that might not work.

2. The prototype should include all features necessary to test your previously defined solution specifications. The prototype should have the functionality to prove that it works as intended in all ways needed to be successful. Functionality should include performance of the robot controls.

3. CAD enables you to plan and check details of how components fit together and move. CAD will give you the opportunity to try different positioning, sizing, and attachments without the expense and time required for a physical robot. Then, when CAD looks good, you can build the prototype more quickly and with fewer problems.

4. Review your prototype to determine if it meets the solution requirements you defined. Ask reviewers to identify any requirements that might be questionable and any that might not be achievable by your prototype.

5. Documentation of prototyping should include a record of the process you used to plan and create a prototype as well as pictures of the actual prototype built. This enables you and others to examine if this step in your design has been done well.

Step 9: Testing Solution Prototype

Testing Your Solution Prototype

With your robot prototype on hand, you now need to see if it meets solution requirements to which it was supposedly built. To do this, plan and conduct tests to verify that these requirements are met.

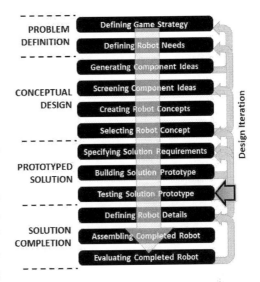

Because you defined requirements for each system, you might want to test system by system. Does the drivetrain meet requirements? Does the collection system meet requirements? Do other systems meet requirements established for them? If you have additional whole-robot requirements, are they met? For each system, spell out questions to be answered, tests to be conducted, and results required for approval.

Take the drivetrain system as an example. Test each solution requirement related to the drivetrain, under conditions like robot matches if possible. If the prototype's weight is not what is expected in the final robot, add missing weight to the robot. Place extra weights in positions that give the prototype a weight distribution expected in the final robot. Since that final weight distribution probably is unknown or it might vary during a match, you ought to test the prototype with different weight balances to see if weight balance affects how well drivetrain requirements are met.

Table 4.4 presents a testing plan for a Mecanum drivetrain. For each requirement question, a testing procedure is outlined, and an approval criterion is set for the requirement. This plan provides important guidance for verifying requirements.

TABLE 4.4. TESTING PLAN FOR ROBOT DRIVETRAIN REQUIREMENTS

Requirement Question	Test Procedure Outline	Required for Approval
Can robot forward speed reach 36 in./s?	Drive robot under full power over 10-ft run; time the sprint	Time less than or equal to 3.6 seconds
Can commanded robot drive hit target within 2 inches?	Use program code to direct robot to drive 10 feet and stop	Error in stop distance less than or equal to 2 inches
Can robot turn about its center 180 degrees in 1 second?	Turn robot 5 complete rotations, and time turn	Time to make turn less than or equal to 10 seconds
Can robot turn 360 degrees within 5 degrees?	Use program code to turn robot 1 complete turn and stop	Error in turn less than or equal to 5 degrees

Can robot strafe sideways at 12 in./s?	Strafe robot at full power across 5 feet, time strafe	Time less than or equal to 5 seconds
Can robot produce 45 lb pull?	Attach robot to anchored force gage, increase power until wheels slip, read maximum force	Maximum force greater than or equal to 45 lb.
Can a motor be replaced in 10 minutes?	Ask team member to replace a drive motor on the robot, time it	Replacement time less than or equal to 10 minutes

Table 4.5 shows sample data for drivetrain prototype testing. Three tests were run for testing each requirement. The acceptance of prototype performance (rightmost column) is based on the mean (average) value of the data compared to the target value. Here we see that three requirements are fully met, three are close, and one is not close. Thus, the drivetrain prototype is not fully meeting requirements.

TABLE 4.5. TESTING RESULTS FOR ROBOT DRIVETRAIN REQUIREMENTS

Requirement	Target	Test 1	Test 2	Test 3	Mean	Accept?
Robot forward speed	36 in./s	37.6	36.2	35.9	36.6	Maybe
Driving accuracy	2 in.	3.6	1.8	2.3	2.6	No
180 deg. turn time	1.0 s	0.9	1.1	0.8	0.9	Yes
Robot turn accuracy	5 deg	6.1	4.3	7	5.8	Maybe
Robot strafing speed	12 in./s	13.7	14.1	14.4	14.1	Yes
Robot pull force	45 lb.	48	46	47	47.0	Yes
Motor replacement time	10 min.	12.2	13.7	9.1	11.7	Maybe

With these results, you need to decide if the drivetrain prototype merits development into the final robot drivetrain, or if this drivetrain requires revision to make it meet the specified requirements. You also have the option of revising the target values set on requirements, which might make the current drivetrain acceptable. Your team should have discussions to determine which action to take.

You will be gathering data on other components of the robot prototype as well. In each case, you gather data, analyze results, and decide if the prototype is performing acceptably. Have serious discussions with teammates and possibly others to decide if you are ready to move ahead. Do not carelessly look at data and accept the results as adequate if the prototype shows serious weaknesses. To move ahead with a flawed design will be costly and demoralizing to your team.

Because this is a critical decision, you ought to hold a design review for the Solution Prototype before proceeding to the next stage of design. A guide for conducting a design review for the Prototyped Solution stage of design is provided in the next section.

Documentation for Testing Solution Prototype

You need to document the process and results of your prototype testing. Both are important to the evaluation of your prototype.

You can document your process by creating a list of tests for each requirement, such as shown in Table 3.4. Your list should include tests for every component of the robot and robot overall to determine its potential for success. You might choose to create a separate table for each component or combine all tests in one table.

Explain how tests were conducted, when, and by whom. Explain conditions surrounding tests that might affect results. You might also document testing with a few photos.

Next, document the test results for every requirement as shown in Table 3.5. Present data for every test run so you can see how much a measurement varies and if it changes systematically over the period of your testing. Then calculate a mean for a measurement, or other parameters that tell you what is necessary to determine if the requirement is met.

Finally, document your conclusions from prototype testing: Is the prototype sufficiently strong to justify advancing to the Solution Completion step? If not, what redesign or rebuilding is suggested? How will your team continue design work and yet meet deadlines? What have you learned that will aid in your design effort?

Resources for Testing Solution Prototype

Additional discussion of testing is found in the last part of the following video. This applies to testing a prototype or a finished robot.

Assembling + Evaluating (40:37 min)   https://youtu.be/d4nZS9v7xjE

Review Questions for Testing Solution Prototype

1. Why is testing a robot prototype important at this point in designing a successful robot?
2. What features need to be built into the robot prototype for testing purposes?
3. What type of test conditions should be used in testing the prototype?
4. What data should be collected in prototype testing?
5. How should the adequacy of the solution design be judged from prototype testing?
6. What should be documented for prototype testing?

ANSWERS TO REVIEW QUESTIONS

1. You need to know if the robot design concept represented in your prototype will performs to meet solution requirements defined earlier. If the prototype does not meet these requirements, you take huge risks to continue robot development using the current robot concept as embodied in the prototype.
2. The prototype needs to have features and functions that will perform what is necessary to evaluate the design against stated requirements. For example, the prototype must be able to collect game elements if any of the solution requirements specify how well the collection is to occur.
3. Prototype testing should be done under conditions like those in competitions. Under game-like conditions, test the prototype's collection, scoring, mobility, and whatever else is specified as requirements.
4. Test every performance multiple times, and record data from every test. Record data in a pre-prepared data sheet with space for recording test conditions (e.g., date, time, participants) and performances (e.g., speed, position, time) related to each requirement being tested.
5. Compare performance data to the corresponding target set by the requirement. For adequacy, see that for each type of performance the target condition is satisfied.
6. Documentation of prototype testing should include descriptions of requirements being tested, test conditions, data collected, results obtained, and interpretations of results. Be sure to document your decisions about next steps in the design process.

Do not exalt yourself in the presence of the king.

Design Review: Prototyped Solution Stage

Before progressing to the Solution Completion stage of design, be convinced that your solution prototype is sound enough to build upon for your competition robot. Your design review must show credible evidence that you have done the following:

- Understood what your robot must be and do
- Specified what features and actions are required for a successful design
- Built a prototype to this set of specifications or requirements
- Proven that the prototype meets these design requirements

Plan a Prototyped Solution design review to address questions such as those above. Invite trustworthy and knowledgeable individuals to examine your work and test its trustworthiness. Ask them to challenge your work, be critical and yet helpful, and offer improvements to make your solution prototype better.

The following are suggested questions that can be used by reviewers to probe your Prototyped Solution work:

- What are your functional design specifications for your robot?
- What are your aesthetic design specifications for your robot?
- What have you built to test achievement of each of these specifications?
- What evidence shows that your specifications have been met by your robot?
- What do you need to do to ensure that your finished robot will be successful?

Ask your reviewers to rate your Prototyped Solution and provide suggestions for improving it. The worksheet on the following page is useful for gathering this feedback.

FEEDBACK FOR PROTOTYPED SOLUTION REVIEW

Review for (team name):

Reviewer names:

Date of review:

Information provided in advance of review:

Please rate the review quality in each area (place X in appropriate column)

Area of Review	Weak	Acceptable	Outstanding	Notes
Sufficiency of functional requirements				
Sufficiency of aesthetic requirements				
Choice of prototyped features & functions				
Quality of prototyped features & functions				
Adequacy of data collection for testing				
Analysis of data collected from tests				
Proof that prototype meets requirements				
Understanding of next steps in design				

Comments on review process:

Suggestions for improving Prototyped Solution:

The end of a thing is better than its beginning.

Chapter 5:
Solution Completion (Stage 4)

Once you have satisfactorily tested your prototyped solution, you are ready to finish building your robot. You have verified that your design concept meets requirements and the systems you have chosen work together. Your design merits the effort necessary to produce an impressive competition robot.

In this chapter you will build your competition robot and validate its adequacy with the people who are the final critics of its success: drive team, other teams, judges, and other people who have a stake in the robot. You need assurance that this robot is a valid solution to the challenges posed by competitions and other events.

Following steps described in this chapter, you will work out the many details required for a completed solution. You will choose parts and raw materials, fabricate and purchase components, and troubleshoot problems that will occur. You will learn new skills, learn how to work under pressure, and learn to prioritize work to meet deadlines. But in the end, you will have a robot in which your team can take pride.

Step 10: Defining Robot Details

Defining Your Robot Details

Once you have proven your prototype successful, you need to develop it into a completed competition robot. What kinds of detail decisions must be made? Some require technical knowledge and calculations, while others are more subjective and a matter of preference.

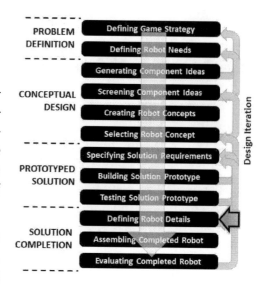

Table 5.1 identifies several types of design decisions often made in completing a robot, along with considerations given in making these decisions.

TABLE 5.1. ROBOT COMPONENT DESIGN DECISIONS AND CONSIDERATIONS

Type of Decision	Considerations in Making Decision
Materials	Weight, strength, flexibility, surface friction
Wheels	Surface speed and distance traveled, surface deformation, traction
Structural members	Loadings, attachment points, size, weight, shape, strength, adjustability
Fasteners	Relative motion allowed, durability, permanence
Motors	Speed and torque output, durability under stalling conditions
Servos	Motion, positioning control
Sensors	Information sensed, software support, response time
Appearance	Color, sheen, shapes, neatness
Control algorithms	Complexity of programming, stability, response time

The following examples of design decision considerations will show you how to approach some of the many decisions ahead.

First you will need to select structural members for your robot. Select them based on the size of robot systems and the robot overall, the loads to be carried, and adjustments that might be required. Plates allow parts to be connected to a single plane. Channels provide varied lengths, bending strength, and protection for wires and cables. Tubes and bars provide good reach with minimal weight. Ideally, you will use materials that are readily available and that do not require specialized cutting, drilling, or bending, which take time and can introduce inaccuracies.

You will need to choose appropriate fasteners to attach parts to one another. Select the attachment method based on relative movement desired between connected parts. Rigid attachment prevents all movement between connected parts, but be aware that connections can loosen under heavy use. Select type of rigid fastener depending on if you want the connection to be permanent or temporary: screws and bolts for temporary connections, rivets and soldering for permanent connections. If you want two pieces to rotate about only one axis, use a hinge attachment. Use a ball joint if you want rotation about a point in any direction. Select a slider if you want movement in one direction along a designated straight line.

Select wheels for desired robot travel, for moving materials in the robot, or for holding moving pieces a constant distance apart. If wheel deformation is important, determine how much deformation occurs for varied loads. If traction is important, measure the friction coefficient to determine the traction force for a given load on the axle. If accurate travel is important, calculate velocity and travel distance from the wheel diameter and rotation, paying close attention to units:

Distance = Pi x Diameter, or **Speed = Pi x Rotational Speed**.

Select a motor based on the speed and torque it must provide. Define the speed and torque required for your load and find a motor that has these characteristics. Figure 5.1 shows the torque vs. speed curves for three motors you might consider. Two of these motors (NR 40 and NR 60) provide the torque needed for the shown load; each motor would operate at the speed where the load line crosses the respective motor curve. With these two options available, you would choose the motor that best meets your desired operating speed for the load. If available motors do not provide the right combination of torque and speed for your load, you might need to place a transmission between motor and load to alter what the motor must provide.

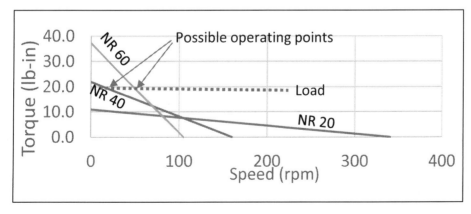

FIGURE 5.1. MOTOR CURVES USED TO SELECT BEST MOTOR FOR LOAD

An important part of detailing your design is determining placement of motors, servos, sensors, batteries, and electronics. Mount motors and servos so they remain solidly in place and yet are easily removed for repairs. Route electrical wires to minimize wire length, especially to motors delivering a lot of power (such as those driving wheels and lifts). Leave slack in wires attached to moving parts so wires are not broken when parts move as intended. Coiled telephone wire as shown in Figure 5.2 allows safe wire extension and pulls in excess length to prevent tangles. Be sure that the wire gauge (diameter) will handle the expected electrical current.

FIGURE 5.2. COILED TELEPHONE WIRE

Whenever possible route wires inside channels or other protected pathways to avoid wire damage during robot operation. To help you in later troubleshooting and wire replacement, label both ends of each wire so you can identify them when they are bundled together.

It is highly recommended to use standard parts whenever they meet your needs for functionality, durability, appearance, weight, etc. If off-the-shelf parts do not meet your needs, you probably must manufacture your own parts or have someone make them for you. Buying custom-made parts can be costly unless you negotiate a donation or discount from the manufacturer.

Making your own parts takes time, skill, and appropriate tools. Even making simple parts probably requires hand tools, a band saw, a drill press, and/or a bench grinder. You might need to train team members in tool use and safety, and realize that quality of the final product might be questionable. A good reminder when making parts is the adage, "Measure twice, cut once."

If you have access to more sophisticated tools, you clearly will need training. Often training videos are available through the company selling the tool. A 3D printer is an additive manufacturing tool; it adds material to build up the shape of part you want. Your CAD drawing software probably has features for converting part drawings into files that drive the 3D printer.

Figure 5.3 shows four robot parts created from a 3D printer. Some are simple holders for batteries and other devices; others are wheels for transferring power. By selecting appropriate printing material and print density, you can obtain the desired strength, weight, and durability of 3D printed parts. Note that 3D printing requires a lot of time.

FIGURE 5.3. ROBOT PARTS MANUFACTURED WITH 3D PRINTER

Precision cutting tools can cut stock (uncut) pieces of metal, plastics, or other materials into precise shapes. Such tools include lathes, routers, and mills. Some machines are run manually by trained operators. Other machines are computer controlled, eliminating the need for a professional machinist; now you must generate computer codes to drive the cutting. Your CAD software might have capabilities to create appropriate files to drive CNC (computer numeric controlled) machines.

Figure 5.4 shows four different parts cut from sheet aluminum and plastic by a CNC machine. Each of the parts was first created as a CAD drawing, then code was developed to drive the CNC machine. Cutting parts with CNC machines is time intensive!

FIGURE 5.4. ROBOT PARTS CUT BY CNC MACHINE

Work of detailing your design is not complete until you have selected all parts and planned their purchase or manufacture. Of course, you need funds for purchases, time for ordering and shipment, and time for training and manufacturing custom parts. Teams often fail to allow time for orders to be processed, out-of-stock parts to become available, or shipping delays. In your planning, allow time for the unexpected!

Documentation for Defining Robot Details

Documentation of robot detailing activity and its results will support robot assembly, and it might be important for detecting errors and avoiding problems ahead.

Document calculations and reasoning used in selecting key parts of your robot. Show your reasoning in selecting motors and gears because this can reveal errors that can be costly. Also show how decisions you made depend upon game or robot conditions, and how the parts used will change if conditions change. These calculations will save your team vital time when changes in the future cause parts to be replaced.

As is done in most construction projects, you should prepare a parts list. In your parts list, show all that is needed to build a component or the entire robot. Table 5.2 is a partial list of parts for a robot chassis. It identifies parts (and part number), the quantity needed, estimated cost of items, and any notes or reminders about the parts. Have a second person check your parts list to be sure nothing is forgotten. Remember to order raw materials, tools, or accessories needed for manufacturing your own parts.

TABLE 5.2. PARTS LIST FOR DETAIL DESIGN OF ROBOT CHASSIS

Part Description	Qty	Cost	Source	Notes
AZ-240 motors for drive wheels	4	$160	XYZ Robotics	Often back ordered
Encoders on motors	4	$80	Know It Electronics	
Mecanum wheels	4	$90	Robot Movers	2 left, 2 right
16-inch 1.2x1.2-inch channels	6	$15	We Make Robots	
16-inch 1-inch extruded beams	8	$18	Squeeze Metals	
1/8-inch bearings (#12-89E)	16	$70	Slippery Sam	
1/8-inch sheet aluminum for side plates supporting wheels	4	$12	Cut per CAD design, using CNC machine	Estimate 4 hours to cut each plate
1/8-inch CNC bits (#67B)	2	$15	Quick Cutz	

After parts lists are prepared for each major system in the robot, you might repackage items into lists for each supplier. This will facilitate ordering.

Resources for Defining Robot Details

Additional discussion of detailing the solution is found in the last part of the video.

   Prototyping and Detailing (30:06 min)   https://youtu.be/9d0DOlFHIEs

A more complete discussion of motor selection is provided in the following video.

   Prototyping + Motor Simulations (37:46 min)   https://youtu.be/ji1U5GMP1gQ

Review Questions for Defining Robot Details

1. Why is detailing the robot solution important before beginning to assemble the final robot?

2. What information should be provided for each part included in a parts list for the robot?

3. How is CAD helpful for preparing a parts list for a robot?

4. Why should you have others review your parts list?

5. For the robot detailing, what should be documented in addition to a parts list?

ANSWERS TO REVIEW QUESTIONS

1. Detailing is a step that pauses work, providing you time to think. The compilation of robot details might help your team to rethink detailing decisions or might reveal some parts previously overlooked. Catching an error now can save much heartache later.

2. Each part should be identified with name or description, part number if available, quantity needed (perhaps allowing for extras), total cost, and source. Check catalogs to be sure your information is correct, and you know what you will order.

3. CAD software enables you to place commercially available parts, devices, and fasteners into your design if they provide parts files. The software might also produce a parts list for you, telling you how many of each part are needed.

4. Having others review your parts list gives you a chance to check your work and identify errors. Both you (as you explain your list) and the reviewer (with fresh eyes and ears) can catch things that make your list better.

5. In addition to your parts list, document your reasoning for selecting parts and devices. Flaws in reasoning might be easier to detect than trying to find errors in the resulting parts list.

Step 11: Assembling Completed Robot

Assembling Your Completed Robot

As you prepare to assemble your competition robot, you probably have many questions and doubts. Will it fit together and look and work as planned?

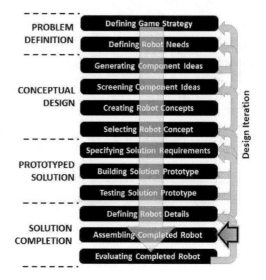

Don't panic. If your prototyping was done thoroughly, the systems should fit, work together, and perform to your solution specifications. If your work to this point has been done well, your risk of major flaws should be low.

Assembling an entire robot is a major undertaking. To speed the work, assign smaller teams to assemble different systems. If you had small teams focusing on individual systems in earlier design steps, you might ask these same teams to assemble their systems. They have the greatest knowledge and most at stake in their respective systems, so they should be able to do the best and fastest work. If any of them struggled earlier, provide them help so they can be quick and successful in their work.

An important challenge in assembling the robot is mustering your workforce and co-ordinating assembly of systems in the robot. Some systems must be ready before others. For example, the drivetrain might be needed before other systems can be attached. Your team's chief engineer or similar person should create an assembly timeline.

As an example, Figure 5.5 shows a timeline for assembling two systems in a robot. This timeline shows assembly steps for only the drivetrain and the collection system, so you will need to add assembly of other systems to this timeline. In the case shown, the collection system is to be assembled then attached to the drivetrain. If the drivetrain assembly falls behind schedule and is not ready when the collection system is to be attached, the chief engineer needs to step in to keep work flowing. The chief engineer might reassign collection system members to aid in assembly of the drivetrain or might ask the collection team to begin testing their system separate from the robot. With good management, assembly of systems can continue and meet the deadline for full robot assembly.

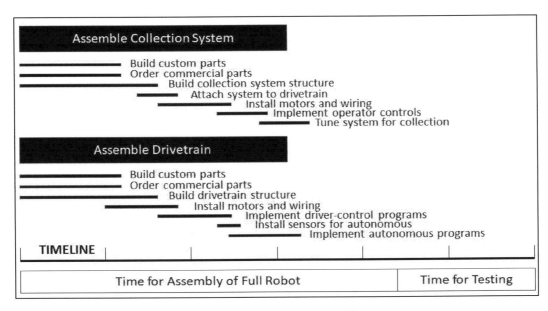

FIGURE 5.5. TIMELINE FOR ROBOT DRIVETRAIN AND COLLECTION ASSEMBLY

The speed of assembly operations will depend upon skills and commitments of people doing the work. A team might need to train novice workers or, better yet, mentor them in earlier stages of the work such as in prototyping. Workers need to know how to use box and open-end wrenches, hex wrenches, and screwdrivers for assembling bolted connections. Grinding and filing might be necessary when fits are tight. Lock nuts and thread adhesives might be used to hold bolted connections in place. The order of assembly is also important, so caution workers to think through the order of assembly for their parts so that fastening is done before access becomes difficult.

An important part of assembly is running wires for powering motors and servos and for transmitting signals from sensors and controllers. Use caution to provide slack in wires so they are not stretched and broken during robot use. Be sure that connections will not be pulled or twisted causing a circuit to be opened, even for an instant, because it might disrupt electronics and require a controller reset in the middle of a match. Not good!

Be careful to avoid conditions that can cause buildup of static electric charges or stray electrical currents generated by wire coils. These can disrupt your electronics. Static electricity is generated when two materials with differing triboelectric values are rubbed together. In your assembly, keep different materials from rubbing a lot, and for materials that must rub (such as materials sliding on floor tiles) choose those materials with similar triboelectric values. When wiring, gather excess wire length in ways that do not create wire coils near other wires carrying high current flow. Coils near high current flow will generate stray current in your wire that is coiled.

Pay attention to robot inspection criteria and game rules that might call for specific assembly details on your robot. Know where shielding is needed to protect you from moving parts that can cut or pinch. Place team numbers and other competition designations where required on your robot. Keep your robot within size and weight limits. Be sure your robot is ready to be tested under actual game conditions.

Documentation for Assembling Completed Robot

The primary documentation for the robot assembly step is photos of the final assembled robot. As noted earlier, photos are needed from many viewpoints and perhaps with parts removed to show key components in place. Label your photos to identify and/or describe functions of important parts.

The assembly process might be documented by photos of team members in action. If certain assembly steps are challenging, capture these with photos or videos so others might learn "best practices." To highlight teamwork and team skills, you might compile a list of annotated assembly activities as shown in Table 5.3. Note that alterations done in assembly are recorded for future reference.

TABLE 5.3. ASSEMBLY WORK SUMMARY

Aspect of Assembly	Assemblers	Notes on Assembly
Drivetrain assembly	Juan, Sara, Soo	Built structure from CAD; wiring and electronics protected using grounding straps
Collection system	Taylor, Jim, Matt	Built structure from CAD; motor changed after design error found
Launching system	Aiko, Abdul, Sam	Built structure from CAD; remade approach to launcher to fix a jam
Wobble goal system	George, Gail, Jorge	Built from CAD; servo gearing added to increase grip on WG

Resources for Assembling Completed Robot

Additional discussion of assembling the solution is found in the first part of the following video:

Assembling and Evaluating (40:37 min)   https://youtu.be/d4nZS9v7xjE

Review Questions for Assembling Completed Robot

1. Why are timelines important for assembling a robot?
2. What training might be required before the assembling process?
3. What actions might be taken to avoid static charge buildup and discharge on your robot?
4. What should be documented for the robot assembly step?

ANSWERS TO REVIEW QUESTIONS

1. The work of assembling a robot is huge, and several parts (systems) need to come together. Timing for combining sections is important, so timelines can help you plan where to put workers and what deadlines need to be met for efficient assembly. Timelines provide a communication tool for the team and a management tool for the project manager or chief engineer.
2. Team members need applicable skills for doing quality assembly. In the least, you all need skill in safely using hand tools and basic shop tools, which might require training.
3. Your team needs to look for places where parts of your robot are rubbing repeatedly with one another or with the field. In these cases, you should attempt to make adjustments that reduce rubbing. If rubbing is unavoidable, select materials so that rubbing parts have similar triboelectric values.
4. To document robot assembly, you might document the process and the assembled robot. You can document the process with an annotated list of assembly actions and photos of key assembly processes. You should thoroughly document your final robot with pictures labeled to identify key features and functionality.

Step 12: Evaluating Completed Robot

Evaluating Your Completed Robot

Now that your robot is assembled, it is ready for critical evaluation. You built your robot based on a conceptual design and solution specifications, defining how it should perform and look. Thus, if everything has gone well, you should have a high performing robot. However, you must yet prove your robot's performance (validate it) in the context of typical matches before you, your team, and others will be convinced.

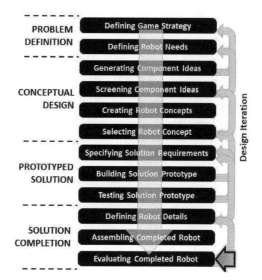

Where should you start robot evaluation? Robot needs defined in design step 2 were big-picture things the robot should do and be for scoring or judging success. These needs, updated by your current understanding of the game and your robot, should be the principal measures of your robot's design success. In contrast, the requirements defined in step 7 were your attempt at defining how robot parts should work, which are less important now. Your evaluation of a completed robot should focus on needs for game scoring and judging success, and design requirements should be tested only if troubleshooting is needed when the game and judging needs are not met.

Table 5.4 defines example robot needs updated from those defined in design step 2. This list includes scoring attempts during autonomous, driver-control, and end game periods. Note that the last row includes a non-scoring need, a need for proven design innovation that will enhance your possibilities for a judged award.

TABLE 5.4. NEEDS TO BE SATISFIED IN COMPLETED ROBOT

Need Short Name	Need Robot Must Satisfy to Meet Team's Expectations
Auto WG Target Zone	Autonomous: Wobble goal is grabbed and delivered to the target zone specified by number of rings on a starter stack (15 pts)
Auto Ring High Goal	Autonomous: From behind a launch line, the robot launches three pre-loaded rings into high goal (12 pts ea.)
Auto Robot Park	Autonomous: The robot is parked on the launch line (5 pts)
DC Ring High Goal	Driver-Control: Collects 3 rings and shoots into high goal (6 pts ea.)
EG WG Drop Zone	End Game: Score WG in drop zone (20 pts)
EG Ring Power Shot	End Game: Score rings in power shot (15 pts ea.)
Robot Transfer Innovation	Robot Innovation for ring transfer from collection to launch performs well enough to prove credible (for judging)

To evaluate robot performance in a simple yet authentic way, test your robot in multiple practice matches. In each match, rate your success in achieving the scoring and innovation performances that you desire. Table 5.5 shows a data collection sheet useful for this testing. Columns 2 to 6 record ratings of each performance in the different test runs (matches). A useful rating scale is 0 = failure to do what is intended, 1 = performs poorly, 2 = almost or close to desired performance, and 3 = performance success as desired. Note here that autonomous WG target zone and park on line and end game drop zone are successful; end game power shot is poor, and others are in between.

TABLE 5.5 ROBOT TESTING FOR GAME SCORING AND INNOVATION

Need Short Name	Ratings for Different Match Runs					Avg Rating
	Run 1	Run 2	Run 3	Run 4	Run 5	
Auto WG in target zone	3	3	2	3	3	2.8
Auto 3 rings in high goal	1	2	3	2	3	2.2
Auto robot parked on line	3	3	3	3	3	3.0
DC 3 rings in high goal (repeats)	2	2	3	3	3	2.6
EG WG in drop zone	3	2	3	3	3	2.8
EG 3 Rings hit power shot	0	1	1	2	1	1.0
Ring Transfer innovation function	1	2	3	2	3	2.2

If you want to know what about a performance is good and what requires fixing, break the performance into crucial actions and rate each on the 0 to 3 scale.

Table 5.6 shows a detailed rating example for the end game power shot. For success, your robot must collect 3 rings, move to proper location, orient toward the first power shot, launch a ring, then repeat orienting and launching for second and third power shots. This data shows that improved power shot performance requires more consistency in orienting the robot toward the power shot and in launching the rings.

TABLE 5.6. EXAMPLE FOR DETAILED DATA ON END GAME POWER SHOT

Desired Action	Run 1	Run 2	Run 3	Run 4	Run 5	Avg	Notes
Collect 3 rings	2	3	2	3	3	2.8	Time too short twice
Drive to proper position	3	3	3	2	3	2.8	Driver error once
Orient to Power Shot	2	3	2	3	1	2.2	Not consistent
Launch trajectory	1	2	2	1	2	1.6	Trajectory varied a lot

Your evaluation of the robot should yield data to identify each problematic action. Use a separate worksheet page, such as the following page, to gather data for each run.

ROBOT EVALUATION WORKSHEET

Recorder:
Persons conducting tests:
Date and Start Time:

Attempt	Desired Robot Action	Rating*	Notes
Auto WG Target Zone (15 pts)	Ring count identifies target zone		
	Grabber grabs, holds WG		
	Robot travels to target zone		
	WG delivered for score		
Auto Ring High Goal (12 pts ea.)	Robot center 74±4 in. from goal		
	Aim ±2 degrees from goal center		
	Rings (3) shot into high goal		
Auto Robot Park (5 pts)	Robot center 60±5 in. from goal		
DC Ring High Goal (6 pts ea.)	Collect 3 rings in 5 seconds		
	Behind line, ±2 deg from goal ctr		
	Rings (3) shot into high goal		
DC Ring High Goal (6 pts ea.)	Collect 3 rings in 5 seconds		
	Behind line, ±2 deg from goal ctr		
	Rings (3) shot into high goal		
DC Ring High Goal (#3)	Collect 3 rings in 5 seconds		
	Behind line, ±2 deg from goal ctr		
	Rings (3) shot into high goal		
DC Ring High Goal (#4)	Collect 3 rings in 5 seconds		
	Behind line, ±2 deg from goal ctr		
	Rings (3) shot into high goal		
DC Ring High Goal (#5)	Collect 3 rings in 5 seconds		
	Behind line, ±2 deg from goal ctr		
	Rings (3) shot into high goal		
End Game: WG drop zone	Grabber retrieves WG in 5 sec		
	Grabber hold WG in transit		
	Arm lifts WG over fence		
	Grabber drops WG in drop zone		
End Game: power shot	Collect 3 rings in 5 sec		
	Adjusts launcher incline for shot		
	Positions ±2 deg for shot		
	Launch scores power shot		
Robot Innovation	Transfer receives rings, indexes rings for transfer, and delivers rings to launcher in order		

*Rating scale: 0 = failure, 1 = poor, 2 = close, 3 = success

Prepare your data sheets in advance so you think through everything needed for collecting match data. In the worksheet shown above, complete everything in advance except the ratings and notes. That ensures completeness of the data.

You will need to run several test matches in which you observe and rate each performance or action. When measuring time for an action, you will need a stopwatch and an ability either to stop action for the measurement or to record the match on video so that you can measure duration of specific actions later.

After completing five to ten test matches, you should have sufficient data to make confident statements about your robot's attainment of stated needs. If serious robot problems occur early in your testing, suspend testing and fix problems before continuing; then gather data for more representative performances.

Table 5.7 shows sample data obtained from five test runs of a robot. Numbers are ratings for each scoring attempt or innovation performance listed in the left column. Rating scale is: 0 = failure, 1 = poor, 2 = close, 3 = success.

TABLE 5.7. SAMPLE PERFORMANCE DATA FOR ROBOT TESTS

Desired Robot Performance	Run 1	Run 2	Run 3	Run 4	Run 5
Auto: WG in Target Zone	3	2	3	0	3
Auto: Rings in High Goal	2	2	3	2	0
Auto: Robot Park on Launch Line	3	3	3	3	3
DC: Rings in High Goal (#1)	2	2	3	2	0
DC: Rings in High Goal (#2)	2	2	2	3	3
DC: Rings in High Goal (#3)	1	2	2	2	2
DC: Rings in High Goal (#4)	3	2	2	1	1
DC: Rings in High Goal (#5)	0	2	0	1	0
EG: WG in Drop Zone	3	3	3	2	3
EG: Rings Hit Power Shot	0	0	1	2	0
Robot Innovation Effectiveness	2	2	3	3	1

What can this data tell you about your robot? Here are some examples:

Question 1: Did the robot meet our stated needs? If success is your goal, simply count the number of 3 ratings for each performance to see if it meets your expectations. If you consider ratings of 3 (success) and 2 (close) adequate for this stage of robot development, you can count the number of performances with ratings of 2 or 3.

Question 2: How consistent is our robot's performance? If you are more concerned about consistency in robot performance, you might calculate means (averages) and standard deviations (variableness) in ratings for each performance. Expect means above 2 and standard deviations smaller than 1 for consistently good performance.

Table 5.8 Shows results of data analysis to address robot performance questions stated above. Can you answer the two questions above for each robot performance?

TABLE 5.8. RESULTS FROM TESTING ROBOT PERFORMANCES

Desired Robot Performance	"success"	"close" or "success"	Mean Rating	Std Dev Rating
Auto: WG in Target Zone	3 (60%)	4 (80%)	2.2	1.3
Auto: Rings in High Goal	1 (20%)	4 (80%)	1.8	1.1
Auto: Robot Park on Launch Line	5 (100%)	5 (100%)	3	0.0
DC: Rings in High Goal (#1)	1 (20%)	4 (80%)	1.8	1.1
DC: Rings in High Goal (#2)	2 (40%)	5 (100%)	2.4	0.5
DC: Rings in High Goal (#3)	0 (0%)	4 (80%)	1.8	0.4
DC: Rings in High Goal (#4)	1 (20%)	3 (60%)	1.8	0.8
DC: Rings in High Goal (#5)	0 (0%)	1 (20%)	0.6	0.9
EG: WG in Drop Zone	4 (80%)	5 (100%)	2.8	0.4
EG: Rings Hit Power Shot	0 (0%)	1 (20%)	0.6	0.9
Robot Innovation Effectiveness	2 (40%)	4 (80%)	2.2	0.8

Answers to Question 1: Did the robot meet our stated needs? If "success" is required, only one performance was 100% successful: Parking on the launch line. Dropping the wobble goal in the drop zone was successful 80% of the time. If being "close" is acceptable at this stage of robot development, eight of the performances met expectations 80% of the time; three did not.

Answers to Question 2: How consistent is our robot's performance? If you expect the mean score to be 2 or above and standard deviation to be less than 1, then five of the performances were consistent. Greatest variability was observed in the autonomous wobble goal delivery to the target zone.

If you are a visual learner, or you want to make a more attractive presentation of results, prepare graphs of data to show the arguments you want to make. Figure 5.6 shows the rating data for the robot test runs, with the ratings of all runs (each run a different shade) stacked in one bar. This graph clearly shows the performance with greatest successes being those with the tallest bars. Shorter bars and those with runs not visible (rating of zero) show performances that rarely were successful.

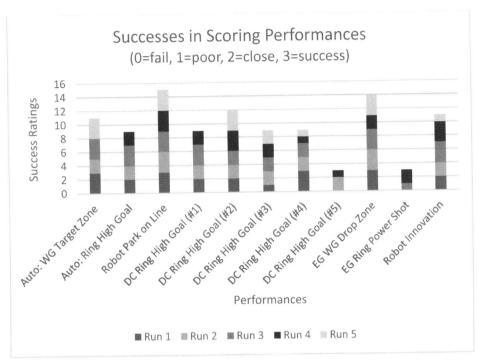

FIGURE 5.6. COMPILED SUCCESS RATINGS FOR EACH TYPE OF ROBOT PERFORMANCE

How should you respond to the results from evaluating your competition robot? First, you need to discuss with your team whether the demonstrated performances are good enough to call for simple fine-tuning of the robot or training of your drive team. If this is all that is required, then your robot is a success. You might make minor robot adjustments to raise the percentage of successes and improve consistency.

If you do not judge your robot to be successful, identify where to focus effort for improvement. Investigate the weaker performances at a deeper level to determine what is "broken." For example, you see (from Table 4.7 above) that the autonomous attempt at scoring rings in the high goal is successful (rating = 3) only 20% of the time and close to success (rating = 2) an additional 60% of the time. This calls for a deeper look at the success of actions making up the attempted scoring.

Probing this performance, Table 5.9 shows ratings of success in three actions that are part of the autonomous scoring of 3 rings in the high goal. Ratings are shown for five different runs. From this data, you see that positioning the robot autonomously at both the proper distance and proper angle from the goal was "successful" only 20% of the time (only run 3), resulting in all 3 rings being scored. Three other runs had "close" positioning but did not score all three rings. Poor positioning (run 5) resulted in missing the goal with all three rings.

TABLE 5.9. SUCCESS RATINGS FOR AUTONOMOUS RING SCORING

Performance	Desired Robot Action	Run 1	Run 2	Run 3	Run 4	Run 5
Auto: 3 Rings in High Goal	Robot center 74±4 in. from goal	2	2	3	2	1
	Aim ±2 degrees from goal center	2	2	3	3	1
	Rings (3) shot into high goal	2	2	3	2	0

This data suggests that you have a problem with autonomous positioning of the robot. Because scoring rings autonomously gets 12 points each, this method of scoring is very important. Improving robot autonomous positioning should be a priority for your team. This improvement might also benefit your power shot scoring capability, which might be added to autonomous and end game periods.

Using similar analysis of the other performance data will probably suggest additional improvements in your robot. When you have a list of such improvement suggestions, you ought to prioritize them so you can focus efforts where they yield the greatest benefit in time available for making improvements.

Once you have targeted redesign on improving specific parts of your robot, consider to which earlier design steps you should go. For example, you might need to add sensors to detect and correct robot positioning errors. This might be seen as a change in the concept for autonomous scoring of rings, then requiring prototyping and testing of these features before incorporation into the competition robot. In your redesign, you will identify appropriate sensors, refine programs to read them, calculate errors, and make position adjustments. This return to earlier steps and making improvements is an example of design iteration. You might iterate for several robot improvements.

Resources for Evaluating Completed Robot

Additional discussion of evaluating the solution is found in the last part of the video.

Assembling and Evaluating (40:37 min)   https://youtu.be/d4nZS9v7xjE

Documentation of Evaluating Completed Robot

What should be documented for this step of evaluating the solution? Of course, you need to document your judgments on how well your robot meets your needs for games and other uses. But you also need to document your basis for these evaluations.

Document your evaluation process by presenting your revised robot needs list and the data targeted to test their achievement, such as shown in Table 5.4. Then identify the number of test matches conducted and show a sample data sheet used for recording data (such as Robot Evaluation Worksheet). If you conducted additional tests to probe weaknesses in performance, explain what these tests included. You might also comment on any specific conditions surrounding the tests, such as identifying the drive team and describing any constraints on the matches.

Document your evaluation results by showing sample data obtained from a series of tests, such as Table 5.7. Then explain and show samples of the analysis of data such as Table 5.8. If you conducted additional tests to probe causes of weaknesses, show samples of this data and its analysis.

Finally, interpret test results in terms of what robot refinement is required or simply desired to make it better. Describe your plans for robot refinement prior to the upcoming competition and perhaps longer term.

This is the time for another design review. You have in hand a finished robot and test data that are great for an outside reviewer to question. The next section describes how to conduct the Solution Completion design review.

Review Questions for Evaluating Completed Robot

1. Why should you evaluate your robot before starting competitions?

2. What can you learn by rating robot successes in practice matches?

3. Why rate how successful, not just success or failure, of performances?

4. If certain targeted performances are weak, how can you determine what needs to be fixed?

ANSWERS TO REVIEW QUESTIONS

1. You evaluate your robot so you know how well it works. Rather than using subjective judgments, gather real data so you have accurate evaluation that enables you to speak to others credibly about what can be expected from your robot. Based on your knowledge of your robot's performance, you can decide where to make improvements and how to "sell" your robot to other teams and judges.

2. Gathering data on robot performances planned for matches tells you how well the robot can perform in matches. You will see which scoring attempts are dependable (successful) and which are not. This will help you think strategically, with data to back up your strategy, regarding how to play the robot game.

3. By rating the degree of success, you will learn which performances are fully successful and which are "close." Knowing that some are close will help you see that small refinements might make these successful. Others that are clearly failures might need major design changes to be successful.

4. For performances that are weak, especially if they are "close," you need to know what ought to be fixed. Dig deeper into robot performances by gathering data on the actions contributing to these performances. Unsuccessful actions need to be fixed. This knowledge will guide you in making robot improvements.

A word fitly spoken is like apples of gold in settings of silver.

Design Review: Solution Completion Stage

Before declaring your robot solution complete, you must be sure that your Solution Completion steps have proven (validated) that your design is sound, and your robot is ready for competition. Before you pronounce your robot "competition-ready," show credible evidence that you have done the following:

    Detailed your robot to meet game, judging, and team expectations

    Assembled your robot with desired form, function, and aesthetics

    Produced a robot that performs with elegance, reliability, and scoring success

    Established credible evidence that validates desired robot performance

Plan a Solution Completion design review to address these questions. Invite one or more individuals to examine your work and judge its excellence. Pick reviewers who have perspectives of the audiences you desire to please: drive team, other teams or coaches, judges, or sponsors. They must be knowledgeable about the robot game and expected competition level and be willing to give you frank feedback. Ask them to be critical and yet helpful in making your final robot solution better.

The following are suggested questions that can be used by reviewers to probe your Solution Completion work:

- What are your robot's needs based on game strategy and judging aspirations?
- What were your greatest challenges when defining details for your robot?
- What are the strongest and weakest aspects of your robot's construction?
- What are the strongest and weakest aspects of your robot's program code?
- What game performance do you want, and what evidence shows you can achieve it?
- What is your greatest concern about your robot, and what do you plan to do about it?

Ask your reviewers to rate your Solution Completion and provide suggestions for improving it. The worksheet on the following page is useful for gathering this feedback.

FEEDBACK FOR SOLUTION COMPLETION REVIEW

Review for (team name):

Reviewer names:

Date of review:

Information provided in advance of review:

Please rate the review quality in each area (place X in appropriate column)

Area of Review	Weak	Acceptable	Outstanding	Notes
Suitability of robot aspirations				
Sufficiency of understanding robot needs				
Sufficiency of defining robot details				
Quality of robot parts and assembly				
Quality of robot code and control				
Adequacy of robot testing process				
Quality of data analysis & interpretation				
Proof that robot meets game needs				
Proof that robot meets judging needs				

Comments on review process:

Suggestions for improving Solution Completion:

SECTION 2: FTC ROBOT DEVELOPMENT JOURNEY

This section discusses how the robot design process is applied when a robot must perform at multiple FTC events, each at a different level of competition. For the first competition, your robot must be viable to compete and have some degree of excellence. At each succeeding competition, your robot again must be viable, but exhibit progressively higher levels of excellence. At many of these competitions you must be prepared to describe your robot development journey to demonstrate your knowledge and communication abilities.

This section contains three chapters:

Chapter 6 guides robot design for the first FTC competition.

Chapter 7 guides robot design improvements between competitions.

Chapter 8 guides you in documenting your robot development journey.

The plans of the diligent lead to plenty, the hasty lead to poverty.

Chapter 6: Designing a Minimum Viable Robot

Creating a competitive robot is not an easy assignment. To make it even more challenging, the robot must be ready for competition in a period as short as 2 months! On top of this, you will compete against ever-stronger opponents as the season progresses. As a design team, one of your greatest challenges is scheduling robot development to fit this dynamic, increasingly difficult season.

The FIRST® Tech Challenge (FTC) season begins in September, and for you it might go as late as April if you are one of the strongest robotics teams. The starting event is a "kickoff" at which the robot game for the season is revealed. The first competition usually is mid-November, allowing your team approximately 2 months to design, build, program, and prepare your robot for a competition event. Later competitions are approximately a month apart to allow teams opportunities to refine and test their robots that continue to be improved. By February, competitions begin to eliminate teams from future competitions, leaving only the best teams advancing to the next level.

Figure 6.1 shows a typical timeline for the FTC season. If this is your schedule, you have 2 months to prepare a basic robot for the first 1-day competition where it will compete in probably 5 matches. Then over the 3 months that follow you will be improving your robot's initial capabilities and adding new ones. You will continue refining your robot, or possibly choose to build a new robot, as long as you are eligible to compete.

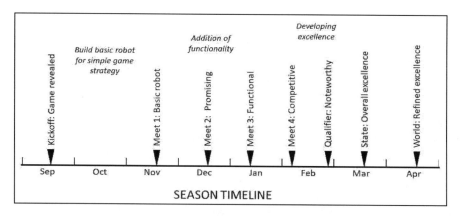

FIGURE 6.1. TYPICAL TIMELINE FOR FTC ROBOTICS SEASON

The big question about your schedule is: How do you fit your robot design process into the season timeline? You do not have the whole season to design a robot. You are designing a robot for your first meet . . . and for your second meet . . . and for your third meet . . . and . . . possibly for the world championship. In this type of season, you are designing one or more robots for different game strategies, different levels of competition, different starting state of your robot, and different capabilities of your team. So, what should you do?

You know with certainty that your team must design a robot for the first meet. And your robot must be able to compete against other teams who probably will not have outstanding robots. So, you need to plan a schedule for this type of design effort. As you are designing and building this robot, you also might want to be planning design of the next robot or robots for later competitions.

What can you do when you have a short time to deliver a working product? In the product development world, when people or companies are eager to get a new product to the market in minimum time, they stress design for efficiency. They focus on identifying "marketable" ideas very quickly, selecting the best one, and getting a "minimum viable product" or MVP rapidly created and tested in their target market. If the MVP proves successful, they use feedback from the test market to drive their next addition of features. By responding to demands of the marketplace, they create an impressive product that over time excels in the marketplace.

Does the minimum viable product approach fit your team's robot development challenge? Can you identify the minimum viable product (robot) needed for your first competition? Would that be a good place to test the direction your robot design has taken? Can you produce a robot with your MVP properties in time to test it in the first competition? Will feedback from the competition guide you toward enhancements that make it a better robot for an overall impressive season?

Whether or not this MVP development model fits the FTC robot challenge, you can adopt some of the MVP approach to get a robot ready for the first competition. You can focus design on achieving only minimum viable robot characteristics for this competition. This means a robot that is successful in a limited number of scoring methods that are crucial to the season. Then after the first competition, you can continue to create a competition robot that will excel in the season ahead.

Table 6.1 offers a schedule for preparing a minimum viable robot in 8 weeks following the kickoff. Using this schedule, your team can rapidly move through the robot design process in time for meet (competition) 1. You must minimize time spent on each design step, and yet use all steps appropriately for producing a minimum viable robot.

TABLE 6.1. SCHEDULE FOR PREPARING MINIMUM VIABLE ROBOT FOR COMPETITION 1

Work Time	Development Desired	Design Steps
Week 1 of Kickoff: extra work sessions to push for definition of MVP in a week	Select robot design concept and have "customers" review it to identify what features they like and give feedback on suggested MVP for competition 1.	1. Defining game strategy for MVP 2. Defining robot needs for MVP 3. Generating component ideas 4. Screening component ideas 5. Creating robot concepts 6. Selecting robot concept
Weeks 2 to 6: meet twice weekly for work sessions	Develop prototypes of components of robot, bring them together, create hardware and controls for MVP robot	7. Specifying solution requirements for MVP 8. Building solution prototype, MVP, to specs 9. Testing solution prototype, MVP, by specs
Weeks 7 and 8: meet twice weekly for work and practice matches	Prepare MVP robot & drive team for competition through practice, refinements, preparation for inspection	10. Defining robot details for competition 11. Assembling completed MVP robot, for competition 12. Evaluating completed MVP robot and drive team through practice, fixes, inspections

The MVP timeline is shown graphically in Figure 6.2. The first week is labelled Sprint, implying that the work will be done very quickly with team members working against a deadline to define the best concept for the MVP robot. The next 5-week period is committed to prototype development and testing, and the final 2-week period is dedicated to preparation for the competition. Note that the MVP prototype must possess viable hardware and software to be ready for the competition.

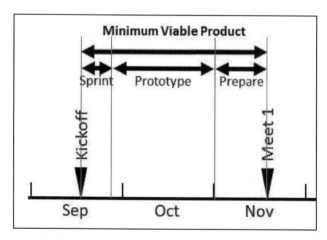

FIGURE 6.2. TIMELINE FOR DEVELOPMENT OF MVP FOR FIRST COMPETITION

This chapter describes the process of MVP design for a competition robot. The next chapter addresses designing a robot after the first competition, for competition 2 and maybe 3 and so on.

With only 2 months to design a minimum viable robot, your team needs to implement a rapid, efficient design process. Use only 1 week for conceptual design of the MVP robot. This will require all team members to be fully committed to an intense week focused on this goal.

Based on Table 6.1, you should complete 6 steps of the design process in this first week. Below are suggestions for accomplishing this feat in 5 partial workdays set aside for your team to get this jump start on your robot's design.

Each of the blocks of time for MVP design is discussed in detail in separate sections below.

Week 1, Day 1: Defining Game Strategy, Minimum Viable Needs

Defining Game Strategy

When the robot game is announced, begin by studying the game, game pieces, and available scoring methods so you can develop a game strategy. (Refer to Chapter 2, design step 1, Defining Game Strategy, to refresh your thinking about game strategy.)

Base your strategy on probable points. Calculate probable points for each scoring method you consider for a minimum viable robot in competition 1. Use Table 6.2 to estimate probable points for each scoring option. In columns 1 and 2, list the options for scoring and points earned by that successful score, for each game period. As a team, estimate the probability of success for one scoring attempt of each scoring option. Then calculate probable points: **Probable Points = Points x Probability**.

TABLE 6.2. CALCULATION OF PROBABLE POINTS FOR COMPETITION 1

Game Period	Scoring Method Options: Competition 1	Points	Probability	Probable Points

Choose a game strategy based on probable points. Tentatively choose two or three or four of these scoring options for your MVP robot. Be sure your team has capabilities necessary to produce this much for the MVP robot in the time available. The result of the Defining Game Strategy step is a list of scoring methods your MVP robot should attempt, with order in which they are used in associated game periods.

Defining Robot Needs

Next determine what your robot must do to successfully achieve your MVP robot strategy. Think through robot actions for each scoring attempt in your strategy and what the robot needs before it can complete these actions.

These actions will define robot needs such as: autonomous sensing of an object, driving to a prescribed location, selectively gathering game pieces, delivering game pieces to a goal, tripping a lever, climbing a ramp, etc. Various needs will demand hardware, software features, operator skill, and/or fabrication capabilities.

Table 6.3 illustrates a process for defining needs for your MVP robot. In each row of column 1, list one scoring method from your strategy. In column 2, list all robot needs, things your robot must do or possess, to carry out the designated scoring method. Note that needs are identified with a loosely named robot component [in brackets].

TABLE 6.3. ROBOT NEEDS FOR SUCCESS IN GAME STRATEGY

Scoring Method	Robot Needs
Autonomous: Move wobble goal (WG) to proper target zone	[Sensor/code] View starter stack of rings; determine if 0, 1, or 4 rings present [WG handler] Autonomously grasp WG; hold WG during transit; release WG [Drivetrain] Autonomously drive from start position to target zone designated by number of rings on starter stack
Autonomous: Launch 3 rings into high goal	[Drivetrain] Autonomously position robot behind launch line aimed at high goal [Ring handler] Autonomously feed rings 1-at-time into launcher; launch ring 5.5 ft at 36 in. height
Driver-Control: Launch 3 rings into high goal	[Drivetrain] Pursue ring on field; drive behind launch line; orient for launch [Ring handler] Pull 3 rings into robot; transfer rings 1-at-time to launcher; launch ring in horizontal orientation 5.5 ft at 36 in. height
End Game: Deliver wobble goal to drop zone	[Drivetrain] Pursue wobble goal; drive to fence by drop zone [WG handler] Grasp wobble goal; hold WG off floor in transit; lift WG over fence; release WG

Define robot needs for every scoring method in your MVP strategy. If meeting these needs is impossible for your team, choose to drop one or more scoring options from your strategy. The outcome from your Defining Robot Needs step is a list of needs your MVP robot must achieve. Ask members to think about the reasonableness of these needs prior to starting the next step.

Week 1, Day 2: Generating Component Ideas

Start Day 2 with a quick review of your MVP robot needs to see if the defined needs are reasonable or any new needs have been identified. Then proceed to design step 3, Generating Component Ideas.

As a team, decompose your robot into major functional components or systems that align with your stated robot needs. Components of your robot might include a drivetrain, a game element collection system, a game element scoring system, and one or more mechanisms for other scoring options of your MVP robot strategy. Based on the size of your team, decompose your robot into a few components that can be assigned to 2- or 3-member sub-teams for generating ideas.

Before generating ideas for robot components, compile the robot needs according to the robot components or systems you defined, as shown in Table 6.4. For example, drivetrain needs shown include autonomous driving and driver-controlled driving to predefined positions on the field or to random positions where rings are found. A wobble goal handler has needs for both autonomous and driver-controlled grabbing, lifting, and releasing a wobble goal. A ring handler needs to collect, singulate, transfer, and launch rings at targets. Note: The MVP robot does not need to gather rings autonomously, but this might be a need for a later robot.

TABLE 6.4. ROBOT NEEDS MATCHED TO ROBOT COMPONENTS

Robot Component	MVP Robot Needs
Drivetrain	Autonomously drive from start position to target zone designated
	Pursue ring on field; drive behind launch line; orient for launch
	Pursue wobble goal; drive to fence by drop zone
Wobble goal (WG) handler	Autonomously grasp WG; hold WG during transit; release WG
	Grasp wobble goal; hold WG off floor in transit; lift WG over 12-in. fence; release WG
Ring handling	Autonomously singulate, feed rings into launcher; launch ring 5.5 ft at 36 in. height
	Pull 3 rings into robot, singulate, transfer to launcher, launch in horizontal orientation 5.5 ft at 36 in. height

With this understanding of robot needs and possible robot systems, begin generating ideas for ways to accomplish stated needs for each component of the robot.

For best results, break your team into small groups to brainstorm ideas for different systems. Generate ideas during short time blocks in which each group is to maximize the number and creativity of their ideas. Use individual brainstorming interspersed with larger group interaction to stimulate additional ideas. A set of idea generation activities and suggested allocations of time are shown in Table 6.5.

TABLE 6.5. PLAN FOR GENERATING NUMEROUS AND CREATIVE IDEAS

Time Block	Idea Generation Activity
10 minutes	Individual: Prepare a list of your ideas for your part of the robot
10 minutes	Sub-team: Share your ideas; identify new ideas that are you generate together
30 minutes	Individual: Conduct online searches for new ideas from other teams, home, industry
30 minutes	Sub-team: Share best ideas with sketches; stretch them for innovation
30 minutes	Team: Share best ideas with one another; pursue ideas for innovation
Homework	Everyone: Ask other people for creative ideas for any part of the robot

When creating ideas, be careful not to criticize or demean ideas because negativity can shut down creativity. Encourage creativity. Be sure that everyone shares their ideas, and that feasible and creative ideas are celebrated. Ask members to post lists and sketches of ideas where every member can see them; this way you preserve ideas and stimulate new ones. As homework, encourage members to search diverse, perhaps unconventional, places for ideas and ask many and diverse people for their ideas. The outcome of the Generating Component Ideas step should be a long list of ideas for each component or system of the robot.

Week 1, Day 3: Screening Component Ideas

To start Day 3, for each robot component, review ideas generated in the previous session and add any new ideas found. Next, perform a quick screening of ideas.

To screen ideas, identify those that are not viable candidates for the component for which they were identified. Set aside any that have these characteristics:

- Cannot provide the functionality required
- Cannot fit in space or weight constraints of the robot
- Requires fabrication or programming skills beyond those of your team
- Not promoted or advocated by anyone on the team

Compile a short list of ideas retained for each component of your MVP robot. Ideally, you will have 2 to 4 viable ideas for each component, and some will be innovative. Provide a sketch or CAD drawing of ideas that are not clearly understood simply by their name.

Week 1, Day 4: Creating Robot Concepts

Next, form viable concepts for your robot. Each concept will include every component (system) and contain one of the generated ideas selected for each component. Look closely at each idea generated for a component to see which ones fit together into a viable robot concept. Table 6.6 shows a way to define and record concepts you generate. As shown, a concept might use a tank drivetrain, a side rotating-brush ring collector, a 2-wheel ring launcher, a 4-bar linkage with finger-type WG grabber, etc.

TABLE 6.6. CRITERIA FOR JUDGING IDEAS FOR DIFFERENT TYPES OF SYSTEMS

Concept Name	Component or System			
	Drivetrain	Collector	Launcher	WG Handler
Example: Tk-SBr-2W-4BF	Tank	Side brushes	2-wheel	4-bar w fingers

Keep this table of concepts handy for the next step, selecting the best concept. As homework, team members might review this set of concepts and see if they believe they are all feasible, and no viable innovative concept has been overlooked. Producing promising new concepts is to be encouraged.

Week 1, Day 5: Selecting Robot Concept, Conceptual Design Review

Selecting Robot Concept

Begin the day with a quick review of the list of robot concepts. See if any can be dropped or any new ones should be added before you begin the selection process.

Next, define criteria for evaluating robot concepts. A suggested list is given below.

- Innovation: One or more features set your robot apart from others
- Speed: Components work together to enable rapid scoring
- Maintenance: Parts require little maintenance and can be repaired quickly
- Cost: Cost of mechanical and electrical parts and fabrication are reasonable
- Feasibility: Team can design, build, program, and operate it for competition 1
- Elegance: Appearance and operation speak of excellence in design

Use a selection matrix as shown in Table 6.7 to rate (1, 3, or 9) each robot concept using criteria weighted (1, 2, or 3) by their importance. Enter concept names as headings for columns. Place ratings in table cells below the corresponding concept and in the row of each criterion. Calculate the product sum of each column to determine a total rating for each concept.

TABLE 6.7. SELECTION MATRIX FOR CHOOSING BEST MVP ROBOT CONCEPT

Criteria	Importance (1, 2, or 3)	Concepts for MTP Robot							
Innovation									
Speed									
Maintenance									
Cost									
Feasibility									
Elegance									
PRODUCT SUM									

Ratings: 1 = weak, 3 = good, 9 = excellent

Once you have determined product sums of ratings, identify the largest one or two, which should point to your best concepts. As a team discuss these ratings to be sure they match your general impressions of the concepts. If ratings do not fit your feelings, ask yourselves, "Why?" If top ratings do not fit your intuition, review your criteria, weights, and ratings to be sure they are justifiable. Make revisions as necessary so that you can defend your selection of top concepts and their order. Finally, use a design review to gain outside perspectives and check your reasoning.

Design Review for Conceptual Design

Conduct a review of your MVP robot design concept development process and outcomes. Seek one or two outsiders knowledgeable about the upcoming robotics competition. Ask them to conduct a conceptual design review at a specified location and time. Invite them in advance so they can study the game and rules. Explain your intent to produce a minimum viable robot for the first competition.

At the time of the design review, conduct the review so that you accomplish your goals in the time available (say 1 hour). A suggested review process is given in Table 6.8.

TABLE 6.8. CONCEPTUAL DESIGN REVIEW FOR MVP ROBOT

Time Block	Design Review Activity
5 minutes	Briefly explain goal of review, robot game, and schedule
10 minutes	Define game strategy and MVP robot needs defined from strategy
10 minutes	Ask for reviewer questions and feedback on robot needs
5 minutes	Describe criteria used for selecting best robot concept
10 minutes	Describe top 3 to 5 concepts and their ranking order
10 minutes	Ask for reviewer questions and feedback on selection process
5 minutes	Ask reviewers for suggestions on improving concept for competition 1 and later
5 minutes	Thank reviewers and ask for completed feedback form with their written review

Keep the review process moving so you get through the final steps. A feedback form is found as the last page of Chapter 3: Conceptual Design.

Results from the design review should give your team suggestions on the robot concept and increase your confidence to proceed with MVP robot design for competition 1. Your review might also give you guidance for the following competitions.

Your selection of a robot concept in 1 week is a major accomplishment that might merit a team celebration. Celebrating this milestone can give you relief after a stressful week and increase your confidence about meeting the next deadline.

Weeks 2 to 6: Building and Testing Prototype MVP Robot

The next 5 weeks focus on developing your conceptual design into a proven prototype robot. First, you define specifications for the prototype. Next, you build the prototype to specifications. Finally, you test and refine the prototype, so it meets specifications. You obtain external review of your prototype to confirm its merits. This proven prototype will be the functional workings of your MVP competition robot.

Specifying Requirements for Prototype

Before building the MVP robot prototype, define specifically what it must do. Here you define how fast, how far, how accurate, how stable, how cool, or whatever the robot must achieve to be successful at the first competition. You probably will define these requirements grouped by the corresponding component or system of your robot. Some examples are shown in Table 6.9 for the ring launcher component of a robot. Most requirements shown are derived from game rules, but the last one is based on the team's game strategy.

TABLE 6.9. EXAMPLE REQUIREMENTS FOR MVP PROTOTYPE

Component/System	Requirement	Target
Ring launcher system	Launched ring height from floor at 65 inches from robot	35.5±2 inches
	Launched ring lateral variation at 65 inches from robot	±2 inches
	Launched ring maximum trajectory height from floor	60 inches
	Launched ring maximum travel in air from robot	190 inches
	Maximum rings on robot at one time	3
	Launch rate for rings (time between rings launched)	2 seconds

This list of requirements must be achieved by the MVP robot and first be proven by the robot prototype. Note that the ring launch requirements are specified for scoring in the high goal. If in future competitions the robot is also to score in the power shot, an additional requirement would be added for launch height: 23.5±2 inches.

Your list of requirements should address all systems in the MVP robot and any that might be for the robot overall. Whole robot requirements might include robot size or weight or weight balance.

Building MVP Robot

Once requirements are defined, build the prototype. Remember that a prototype is built simply but must have capabilities to demonstrate achievement of requirements. Thus, you probably will build your prototype from standard beams, channels, rods, gears, sliders, etc. Begin by building each system of your concept; then join systems.

Be sure that each system is constructed with the movements and rigidity needed to produce smooth and accurate performances of requirements. Ideally, you should create a CAD of a prototype system to define exact sizes of parts and desired movements. At this point it is ok to guess at parts that will be strong enough; more exact sizing can come later. Assemble parts so they can be easily accessed and replaced if necessary.

As you build individual systems, test them to see that they perform generally as expected. Does the collection system work? Does the launcher work? Does the wobble goal grabber work? Now is the time to fix obvious problems before the systems are integrated into the full prototype. More complete testing will occur after the entire prototype is constructed.

As you build each system, also develop the control software to give you whatever control is necessary for testing your prototype. This might require capabilities for manually powering systems, reading sensor values, varying speeds of mechanisms, etc. You probably do not need sophisticated control of robot functions in your MVP robot, so this level of programming is not required for your prototype.

Testing MVP Robot Prototype

Once the prototype is assembled, you are ready to test it. Begin by examining your list of requirements and sorting them into groups that can be tested together. For example, requirements related to the ring collection system can be tested during a series of tests with the collection system. Similarly, launching system requirements can be tested together. Plan the series of tests so they enable you to efficiently collect data that evaluates achievement of these requirements.

Develop a test plan as illustrated in Table 6.10 for testing the ring launcher requirements. Multiple requirements are evaluated in each of two tests defined. The first test measures variations in ring trajectory, both height and lateral variations. The second test measures ring trajectories (maximum height and distance) as well as time to launch 3 rings.

TABLE 6.10. TEST PLAN FOR MVP PROTOTYPE

Requirement	Testing Procedure	Data Collected
Launched ring height and lateral variation at 65 in from robot	Launch 3 rings in succession 65 in. from target 35.5 in. from floor. Repeat launch sequence 10 times.	• Height error of ring striking target • Lateral error of ring striking target
Launched ring trajectory limits and launch frequency	Launch 3 rings in succession (as if aimed at high goal) but pointed across open floor. Repeat 10 times.	• Maximum height of ring trajectory • Distance where ring strikes floor • Time to launch 3 rings

Define and conduct tests for evaluating achievement of requirements for each system and for the robot as a whole. If any requirements are not achieved, discuss the significance of these deficiencies, and determine what should be done. If results are close to requirements, you might assume that deficiencies can be resolved in the finished robot. If deficiencies are serious, you probably need to fix related problems in your prototype and test again.

You need to be confident in your prototype before moving on. This is the time for another design review to get expert advice on your prototype before proceeding.

Design Review for MVP Robot Prototype

As done earlier, gather external reviewers to critically examine the merits of your prototype for further development. It is critical that weaknesses are detected here so they do not plague your design from this point forward.

Define and follow a design review procedure as discussed earlier: preparing reviewers to review an MVP robot, describing your processes of MVP robot development and testing, and concisely presenting your prototype and evidence of its achieving requirements. Invite reviewers to do a probing review and provide suggestions for improvement. They might pinpoint improvements they feel are necessary before you are ready to proceed. Ask reviewers to complete the feedback form provided at the end of Chapter 4, Prototyped Solution.

Weeks 7 to 8: Preparing and Testing Competition MVP Robot

When the prototype has been proven, begin developing your MVP prototype into a competition-ready robot. But now also is the time to realize that everything lacking in your MVP robot probably cannot be completed before the competition. Some things will be left undone in the MVP robot. You need to prioritize what gets finished and prepare wisely.

Defining MVP Robot Details

Make work prioritizations based on how vital something is to the competition and how likely you can complete it with resources available. Table 6.11 illustrates how you can prioritize tasks by calculating a priority rating for each task. Assign each task a number that rates its importance to the upcoming competition, with 5 = required to 1 = unimportant. Assign another rating for the likelihood you can successfully complete the task in time, with 5 = guaranteed to 1 = extremely doubtful. The product of these two ratings yields a priority rating, with the larger rating being higher priority.

TABLE 6.11. EXAMPLE PRIORITIZATION OF TASKS FOR COMPLETION PRIOR TO COMPETITION

Task	Resources Needed	Importance to MVP	Likely Success	Priority Rating
Stencil team numbers on robot	Stencil, 2 hr	5	5	25
Resize motor to increase lift speed	Calculations, 1 hr	2	4	8
Replace lift motor and gears	New parts, 3 hr	2	4	8
Fix jamming in collection system	New ideas, 5 hr	4	3	12
Fix errors in launch accuracy	New ideas, 5 hr	2	3	6
Add power shot capability	Adjust launch, 6 hr	1	2	2
3D print battery holder	3D cartridge, 5 hr	1	4	4
Practice matches	15 min./match	4	4	16

From this table you see that you shall place team numbers on the robot. You also must do practice matches to improve scoring, to obtain data to know how well your robot performs, and to identify problems that can be fixed before the competition. Another priority is fixing a jamming problem in the collection system because it prevents scoring and probably can be fixed.

Determine a work schedule based on priority of the task and order in which tasks make sense. For example, adding numbers to the robot can be done any time after robot shielding is ready. Many tasks can be done in parallel if work is distributed among team members. Practice matches can be run simultaneously with gathering test data.

Part of your preparation for completing the robot is selection of parts (sizing motors, choosing beam sizes, selecting gear sizes, etc.) and determining placement on the robot. CAD drawings can help with spacing questions. Choosing gears, pulleys, chains, and other drive elements can be done by considering speed and torque requirements for wheels and shafts. Refer to references such as *Pre-Engineering Primer, 2nd Edition*, for guidance on making these selections.

Assembling MVP Robot

Assembling the MVP robot from the prototype must be accomplished with help from the entire team to complete it in a week. You probably will have needs for several minor additions or substitutions to the prototype, and hopefully, members can simultaneously work on different sections of the robot. Oftentimes, small sub-teams will need to schedule work at different times to avoid conflicts. Your team's engineering manager will need to coordinate work to get the most done with resources available.

Robot preparation for competition also requires fine tuning of your programs. Be sure you have followed competition rules for robot initialization on the field, starting autonomous programs, and operator-controlled robot performance. Consider how much capability you have time to enable for varied game conditions. For the first competition, you can operationalize only a subset of what you would like in later competitions.

During your assembly, pay attention to details. Be sure you fasten parts so they do not work loose in a match. Make joints so they limit movement as planned and do not wobble; this requires careful drilling, grinding, bending, and use of high-quality hinges, pins, bearings, and sliders. Mentor team members in good, safe manufacturing processes. Test your robot's joints, movements, power, and controls as they are being developed and installed; catching and fixing problems is easier while components are simple.

Evaluating Completed MVP Robot

In the short time available for finishing and testing the MVP robot, testing must focus on the most vital issues. You need to know that your robot performs functions necessary for scoring you planned in your game strategy. If time is available, you also would like to know how well it performs each scoring function.

Define a testing plan for the MVP robot that can be done as your drive team runs practice matches. Table 6.12 shows a sample data sheet useful for answering the question about robot scoring effectiveness. Mark each time a scoring attempt is made and when a successful score is made. Effectiveness can be calculated later: number of successes divided by number of attempts. If a score is made accidently, it does not count in this data sheet. Record as comments any unusual conditions or reasons for missing a scoring attempt.

TABLE 6.12. SAMPLE DATA SHEET FOR EVALUATING MVP ROBOT

Scoring Method	Attempts	Successes	Effectiveness	Comments
Autonomous: Move wobble goal to target zone				
Autonomous: Launch rings into high goal				
Driver-Control: Launch rings into high goal				
End Game: Deliver wobble goal to drop zone				
End Game: Launch rings into high goal				

Prepare a summary of your testing results and conclusions. In it you might state that certain scoring methods are well developed while certain others need improvement. You might also list common causes of scoring failures to guide future design changes.

Preparing for First Competition

Other last-minute preparations for your first competition should be completed during this last two weeks prior to the competition. Preparations might include packing items needed at the competition and planning team interactions there.

Items to be packed might include:

- Spare parts for your robot's most vulnerable components
- Tools for repairing and adjusting your robot
- Batteries, chargers, test instruments
- Team brochures and giveaways
- Photographic and video equipment

Plan for interactions with teams, coaches, and others that help you understand the dynamics of the competition and what a team must do to advance through later competitions. Get acquainted with other teams and learn about their designs and design philosophies. Note the best robots and seek to understand how their robots perform.

Gather information to guide you in future development of your robot. Ask teams for their opinions on your robot. Gather match data for your team and others to understand the competition and how well your robot design is meeting demands of these competitions. Ask teams and coaches what they want to see in robots that partner with them in the future. What do they envision for the best robots in the world?

Look for opportunities to help other teams that are struggling. And most important of all, be gracious and have fun!

With the completion of your first competition, you face new challenges in preparing for the next competitions. The next chapter addresses this period in your robotics season.

You do not know what things sowed will prosper, so try many.

Chapter 7:
Designing a Robot of Excellence

A robot that achieves a season of excellence requires excellence in design every step of the way. Whether you created a minimum viable robot first or set your eyes on building a robot of excellence from the start, you need to move forward from your starting point with excellence being your focus.

If you created a minimum viable robot for the first competition, you should have gained valuable information for starting the robot of excellence. In this chapter you will be challenged with questions to ponder, and you will be shown advanced tools to apply to build excellence into each design step.

A major challenge for the season is producing intermediate robots that are viable and continually building on excellence. Between competitions, you must carefully choose improvements that can be completed in available time and that make your robot competitive in the upcoming competition. Therefore, immediately after the first competition (or at the start of the season) you should tentatively define a robot development plan for the remainder of the season. Then begin to implement the plan for the upcoming competition. Then after each succeeding competition, review your plan and make revisions based on what you learned in your recent competition experience.

Defining Game Strategy

How do you develop a long-term plan that produces viable intermediate robots? You need a game strategy that evolves during the season to fit the stages of your robot's development. Your strategy should increasingly attempt additional or more challenging scoring attempts. These higher aspirations will call for added features on your robot or refinements that increase the speed and effectiveness of scoring attempts. Your challenge is to identify inspiring and yet achievable improvements for each successive competition.

First, develop a game strategy for your final robot of excellence. What do you envision your best efforts in design, building, program control, and drive team performance could produce by the end of the season? List every scoring method you might use and update your estimated success probabilities and times to score for the season's end.

Table 7.1 provides an example for developing a long-term game strategy. Scoring methods being considered, ones we imagine our robot will eventually achieve, are listed in column 1. Column 2 shows the number of points earned from successful attempts with the scoring method. Column 3 shows your estimates for time required for that scoring method. Column 4 shows your estimates of probability that the entire scoring method is successful. Probable points (average points expected) are calculated: **Probable Points = Points x Probability.** Scoring efficiency, probable points per unit time, is calculated: **Efficiency = Probable Points / Time.**

TABLE 7.1. EXAMPLE GAME POINT ANALYSIS FOR STRATEGY PLANNING

Scoring Method Options	Points	Time to Score	Probability of Success	Probable Points	Scoring Efficiency
Auto: Wobble goal to target zone	15	8	0.90	13.5	1.7
Auto: 3 rings into high goal	36	7	0.85	30.6	4.4
Auto: 3 rings hit power shot	45	10	0.75	33.7	3.4
Auto: Park robot on launch line	5	3	0.95	4.8	1.6
Driver: 3 rings into high goal	18	9	0.85	15.3	1.7
End: 3 rings into high goal	18	9	0.85	15.3	1.7
End: 3 rings hit power shot	45	12	0.75	33.7	2.8
End: Wobble goal to drop zone	20	7	0.95	19.0	2.7

From this information, you can define a game strategy for your envisioned final robot of excellence. Let us consider data for each game period indicated.

For the 30-second autonomous period, the largest number of points can be expected from scoring 3 rings in high goal or hitting power shot: one or the other, not both. The wobble goal and parking options can be completed in 11 seconds, leaving 19 seconds for scoring rings, which is ample for either high goal or power shot scoring option. You might choose the power shot option because it offers more probable points. If your team does not envision that your robot can collect additional rings autonomously, then the autonomous strategy is: take wobble goal to target zone, score power shot, and park on launch line. This yields on average 52 points in 21 seconds.

During the 90-second driver-controlled period, scoring is accomplished through collecting and scoring rings into the high goal. Based on estimated time, 10 attempts can be made, and 10 x 15.3 = 153 average points are expected.

During the 30-second end game period, point efficiency is important because you are not limited by the number of scoring attempts in ring scoring. Thus, you might choose power shot scoring over high goal scoring. Your strategy might include wobble goal to drop zone (taking 7 seconds) and then one attempt at 3-ring power shot (12 seconds), leaving 11 seconds. This allows a choice of one 3-ring high goal netting 15 points or one 2-ring power shot netting 22 points., which might be better.

This game strategy offers 52 + 153 + 75 = 280 points as the probable score. It also requires the robot to score both wobble goal methods, high goal under driver control, and both autonomous and driver-controlled power shot scoring. Your design team must be able to deliver these capabilities with the scoring times and probabilities indicated earlier.

If you are not satisfied with this potential score, then you might consider either increasing scoring efficiency or adding scoring options to your strategy. If you believe you can add autonomous ring scoring capabilities, then the autonomous score might be increased for your final robot of excellence. You might increase scoring efficiency by automating robot actions to make them faster; this might include automatic alignment for collecting and/or scoring rings. Your vision for excellence can grow as you gain capabilities in your team.

Document the strategies you defined and rationale for differing strategies for different competitions.

Defining Robot Needs

Your robot needs to possess the scoring capabilities defined in your game strategy. Your challenge is to define those needs for each competition ahead. In distributing these needs, remember that your team must have the time, abilities, and other resources to develop the robot to meet the needs defined for any given competition.

An important step for planning is to create a table of robot needs for different points in the season. Begin with the needs you just defined for your robot of excellence (say, for the state competition) and recall the needs you defined for the MVP (first competition) robot. Then fill in needs at intermediate competitions so they provide a continuum from season start to finish.

Table 7.2 provides an example of needs defined for a robot at start, a midpoint (qualifier), and end of the season (state). Note that for the first competition, the power shot scoring method is not part of the strategy. It is shown at the qualifier and state-level competitions. Also note that for each need a percent success is targeted, intentionally leading to the desired success in the robot of excellence (state).

TABLE. 7.2. ROBOT NEEDS AT DIFFERENT POINTS IN SEASON

Scoring Method	Competition 1	Qualifier	State
Auto: WG to target zone	Rings read 80% WG delivered 80%	Rings read 90% WG delivered 90%	Rings read 95% WG delivered 90%
Auto: 3 rings into high goal	Robot positioned 80% Rings hit goal 70%	Robot positioned 85% Rings hit goal 75%	Auto navigation 95% Robot positioned 95% Rings hit goal 85%
Auto: 3 rings hit power shot		Robot positioned 80% Rings hit goal 70%	Auto navigation 95% Robot positioned 95% Rings hit goal 75%
Auto: Park	Park on line 80%	Park on line 90%	Park on line 95%
Driver/End: 3 rings into high goal	Rings collected 70% Robot positioned 80% Rings hit goal 70%	Rings collected 80% Robot positioned 85% Rings hit goal 75%	Rings collected 90% Auto navigation 95% Robot positioned 95% Rings hit goal 85%
End: 3 rings hit power shot		Rings collected 80% Robot positioned 85% Rings hit power shot 60%	Rings collected 90% Auto navigation 95% Robot positioned 90% Rings hit goal 75%
End WG to drop zone	WG retrieved 80% WG delivered 85%	WG retrieved 90% WG delivered 90%	WG retrieved 95% WG delivered 95%

In addition to needs identified above associated with planned improvements, your team might have needs that resulted from poor performances at the previous competition. Thus, you will have several improvements desired before the next competition.

If the next competition is only 3 weeks away, you must prioritize improvements. One way to do this is by calculating a benefit rating associated with fixing a problem or addressing a need. You quantify the seriousness of potential failure in the current state, likelihood the failure will occur, and the drive team's ability to prevent the failure (controllability) from occurring during a match.

Table 7.3 provides an example of deriving a benefit rating for improvements considered after the first competition. Seriousness of related failures is rated from 1 = no effect to 5 = kills robot. Likelihood of the failure is rated 1 = very unlikely to 5 = almost certain to occur. Controllability is rated 1 = very easy to prevent to 5 = nearly impossible to prevent the failure.

TABLE 7.3. RISK ANALYSIS OF ROBOT FIX-UP OPTIONS

Improvement	Seriousness	Likelihood	Controllability	Benefit Rating
Improve miscounting of rings in starter stack	3	2	5	30
Prevent erred travel to target zone	3	2	4	24
Prevent erred auto robot positioning for ring shots	3	3	5	45
Prevent erred driver robot positioning for ring shots	4	3	3	36
Prevent jamming of launcher ring feed	4	4	4	64
Prevent inconsistent ring launching	4	3	4	48
Prevent collector inability to get rings in corners	2	2	3	12
Prevent ring stalling in collection system	2	3	3	18
Prevent current inability to score power shots	2	5	5	50
Prevent misses in retrieving wobble goal	3	2	2	12

From this table, you see from benefit ratings that the greatest benefit (largest benefit ratings) can be gained by preventing jamming of the launcher ring feed. This problem prevents ring scoring, which greatly reduces your team's ability to score points. Fixing the errors in robot positioning and inconsistency in ring launching also can significantly improve scoring potential. Your use of navigation targets to automate robot positioning is shown in your strategy as a need in a later competition, so you should ignore its benefits here. These benefit rankings suggest that improving ring scoring should be top priority now.

Developing capability to score power shots promises additional benefits, but this capability is not identified as a need until later competitions. Thus, it might become top priority for a later competition.

Note that your team can be making other refinements to your robot to add excellence in appearance and improved performance if these do not require major team effort. With such a short time before the next competition, focus your energy on what yields the most benefit and can be completed with excellence.

To be sure you have defined the problem adequately, you should seek a quick design review by someone outside the team. Simply describing your process and decision will provide an opportunity for another person to detect errors in thinking. Now is the time to catch these mistakes.

With top priority set for fixing the launching of rings, you are ready to move to the next design step: generating ideas. Be sure to document your decision and the basis for this decision.

Generating Component Ideas

It appears that your robot's launching system is inconsistent in launching and rings can jam during entry to the launcher. You have chosen the 2-wheel launcher concept for ring launching, so ideally you should stay with this concept and improve its details. Your goal is to achieve excellence in this launcher, so the ring feeding system must work with excellence.

Study carefully the dynamics of your robot launching system. You might make video recordings of the ring entry and ring exit motions. You probably will see that the ring's approach varies, which affects the entry point between the 2 launcher wheels, affecting the launch speed and orientation. Problem: The launcher needs a consistent feed.

Begin idea generation for a ring feeder into a 2-wheel ring launcher. Be sure to generate ideas and not grab the first good idea that comes along. Look outside your team for ideas. For example, look at assembly lines in which cylindrical objects are positioned for the next action. This might include jars and cans being filled. Compile and categorize your ideas. Table 7.4. shows examples of ideas that might be generated, sorted by category.

TABLE 7.4 IDEAS FOR RING FEEDER INTO A RING LAUNCHER

Category	Ideas for Ring Feeder
Passive – stand alone	Slide into narrowing opening Rolling down incline, then tip flat
Active part of transfer	Finger on transfer device to push ring out Transfer triggers finger to snap ring away
Active – stand alone	Rotating overhead wheel to pull ring toward launcher Rotating overhead fingers to push ring toward launcher Overhead ratcheted finger to push ring only in one direction Reciprocating overhead arm to push ring into launcher Rotating overhead brush to brush ring toward launcher Rotating pair of wheels at sides of ring to push ring to launcher Robotic arm grabs and pushes ring into launcher

Before moving on, ask yourself, "Have we found ideas that are innovative and feasible and will build excellence into our robot?" If not, spend time seeking better ideas. Remember to seek excellence in every improvement or addition to your robot.

Screening Component Ideas

Next, you will screen your ideas for a ring delivery system to the launcher. Since you have already selected the launcher concept, you ought to make this a refined selection of the one best idea that fits with the launcher concept.

So, what are the criteria you should use for this selection? Remember that excellence is your goal. Suitable criteria might be consistency in ring delivery, innovation, not interfering with ring launch, fit in available space, elegance in design, feasibility in manufacturing, and cost.

You can rate each idea using your criteria, as shown in Table 7.5. Since you are seeking excellence, use weighted (1, 2, 3) criteria to consider the value of each. Rate ideas with values of 1, 3, and 9 to increase separation among ideas. The products of weight and rating are summed in each column to yield a weighted total for each idea.

TABLE 7.5. SCREENING RING FEEDER IDEAS FOR EXCELLENCE

Criteria	Weight	Narrowing slide	Rolling incline	Push from transfer	Throw from transfer	Overhead wheel	Overhead fingers	Ratcheted finger	Reciprocating arm	Overhead brush	Wheels to side	Robotic arm grabber
Delivery consistency	3	3	3	1	1	3	3	9	9	3	3	1
Innovation	2	1	9	1	3	1	3	9	3	1	1	9
No interference	3	3	3	3	3	1	1	9	3	1	3	1
Fit in space	3	3	3	3	3	1	1	3	9	1	3	1
Design elegance	2	3	9	1	3	1	3	3	3	1	3	3
Feasibility	3	9	3	9	3	3	3	3	3	3	3	1
Cost	1	9	9	9	9	9	3	3	3	3	3	1
Weighted Total		71	81	61	51	37	39	99	87	31	47	37

From your totals, pick the top-rated ideas. Next, integrate these ideas with the remainder of the launcher system to see how well they fit together.

Creating System Concepts

Your challenge is to see how well your top-rated ring feeder ideas fit into a launcher system so that they bring excellence to your robot. This is the time to sketch or prototype your top ideas in conjunction with launcher wheels to see how they fit together. You need a way to determine more convincingly that your ideas will fit with the launcher system and meet the criteria you chose.

As an example, look at the top-rated idea of a ratcheted finger. Figure 7.1 shows a sketch of how this might be implemented. Pay attention to how the rings will be aligned so they enter the launcher along the centerline between the wheels every time. Also consider how the mechanism fits inside allowable space. Ask if it might possess innovative elements, such as nonlinear springs, sculpted finger to optimize ring positioning, etc.? Look for ways to gain excellence in this part of your robot.

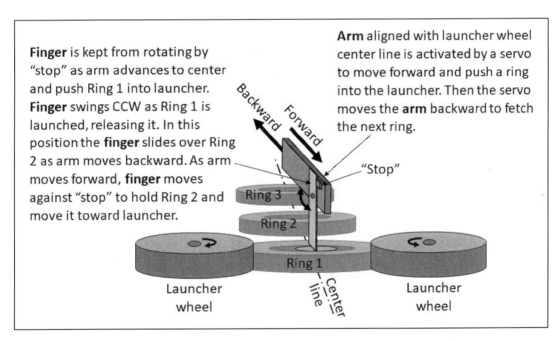

FIGURE 7.1. CONCEPT OF RATCHETING FINGER IN 2-WHEEL LAUNCHER

You will want to examine each of your top ideas in the context of the launching wheels. If none of them seems to fit well, you might need to tweak an idea or alter the launching system in conjunction with the ring feeder ideas to get a better system concept.

When you are ready, move to the next step.

Selecting Robot Concept

Your next decision is critical to producing excellence in the system you chose to improve. You now need to confirm your best improvement idea, as part of this system.

Use a Pugh matrix to rate your proposed concepts for the improved system. Note that your proposed concepts are more specific than the ideas you screened earlier, because these are understood as part of the system. Because you know more about the ways the system must work to be excellent, you might also choose to refine selection criteria or weights of criteria.

Table 7.6 shows an example Pugh matrix for evaluating four concepts. Note that these concepts are refined from their initial ideas, in some cases adding features that promise improvement. Some criteria are expanded and/or clarified to better fit excellence.

TABLE 7.6. PUGH MATRIX FOR IMPROVED LAUNCH SYSTEM CONCEPTS

Criteria	Weight	Servo-actuated arm with ratcheted finger	Same as previous plus tapered chute	Rotating wheel with ratcheting fingers	Same as previous plus tapered chute
Ring placement consistency	3	3	9	3	9
Ring delivery speed	2	3	3	3	3
Innovation	2	3	3	3	3
Fully releases ring to launcher	3	9	9	3	3
Fit in space	3	3	3	1	1
Design elegance	2	9	9	3	3
Feasibility	3	3	3	3	3
Cost	1	3	3	3	3
Weighted Total		**87**	**105**	**51**	**69**

Results of this evaluation suggest that a servo-actuated arm with ratcheted finger and tapered chute will produce the greatest excellence. It offers accurate placement, clear release of rings, and opportunities for design elegance—elements of excellence.

Now would be a good time to seek independent review of your concept selection. A quick design review might catch any mistakes and save you many tears later.

Specifying Solution Requirements

Because your between-competition design efforts are focused on building excellence in one or more components of your robot, focus your specification of requirements on systems being improved. Define specifications that clarify excellence for these systems.

For example, look at requirements for the ring launching system discussed in the previous sections. Table 7.7 presents requirements for functionality and features of the ring launching system. Note requirements specified for trajectory (height and lateral error at target), frequency of launching (time between successive rings), jamming (occurrences of jamming) at entry to the launcher, and physical extremity of system.

TABLE 7.7. SOLUTION REQUIREMENTS FOR RING LAUNCHING SYSTEM

Scoring Method	Launcher Requirements	Value
3 rings in high goal	Launched ring height at goal	35.5 ±2 in.
	Launched ring lateral error from center of goal	±3 in.
3 rings hit power shot	Launched ring height at target	35.5 ±2 in.
	Launched ring lateral error from center of target	±2 in.
Launch speed	Time lapse between successive ring launches	1 s
Launcher jamming	Rings jam at entry to launcher	< 2%
Launcher height	Height of top of launcher system from floor	17.5 in.

Create a set of requirements for every system or feature being improved. This forces you to define what constitutes excellence, and it holds your team accountable for producing excellence in each improvement.

Once specification of requirements is completed, proceed to the next step in the design process.

Building Solution Prototype

Now that requirements are defined, build a prototype of the improved system so that performance can be tested. Your prototype should have all the functionality essential to achieve the desired performance. But it need not have some of the finery expected for excellence.

For example, if the launcher improvement is a new feeder system for rings, build this feeder system as it will be incorporated into the launcher itself. The selected concept has a servo-actuated reciprocating arm that pushes a ratcheted finger that drives the ring into the launcher wheels. The finger, by rotating freely, releases the ring as it is launched. When the arm is retracted, the finger passes over the front of the next ring then drops down to push this ring for the next launch. A tapered chute (space between two fences) forces the ring to enter the launcher centered between the two Launcher wheels. Figure 7.7 shows the principal parts of the prototype.

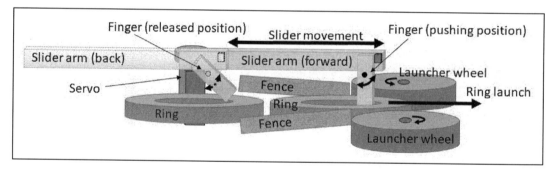

FIGURE 7.7 RING FEEDER PROTOTYPE FOR RING LAUNCHING SYSTEM

The prototype might have some of its parts made adjustable until precise positioning is determined. For example, the travel distance of the servo-controlled arm might be adjustable. The taper (narrowing) of distance between the two fences leading to the launcher wheels might be adjustable. The length and shape of the finger might need to be varied to see what works best.

Build this prototype so it can be tested against the requirements specified earlier.

Testing Solution Prototype

Now that your prototype is built, test how your design change has achieved excellence desired. Develop a testing plan to gather data on launcher performance as defined by your solution requirements. Think also about how you might vary test conditions to gain added useful information.

As an example, consider how you might test the launcher requirements listed in Table 7.6. A testing plan for these requirements is shown in Table 7.8. Note that testing the launcher accuracy can be accomplished by one set of tests, using the same conditions for repeated launches at the target. However, additional sets of these tests might be needed for testing different configurations of the launcher parts. Another series of tests for testing the speed of launches can be done along with observations of launcher jamming. Again, you might want to change launcher configurations and repeat this test series to see what configurations work best.

TABLE 7.8. TESTING PLAN FOR RING FEEDER AND LAUNCHING SYSTEM

Launcher Requirement	Launcher Requirements	Success
Launcher accuracy at 5.5 ft away	Launch at target 5.5 ft away, 36 in. high Measure location of hit from target	Error <= ±2 in. to side Error <= ±2 in. vertical
Launcher operation firing 3-rings in succession	Launch 3-ring bursts many different times Measure time between launches Record instances of jamming	Lapse time <= 1 s Jams < 5% of time
Height for top of launcher	Measure launcher height after every adjustment for different test configurations	Height <= 17.5 in.

After running these tests, you should be ready to identify the best configuration for the launcher parts. The best fence spacing, slider arm placement and movement, and finger shape can be determined from the data. This will give you solid information for a launcher design that delivers excellence.

Note that you also can test any other design changes being considered. When testing, do not make a quick "It works!" conclusion, but gather data to show how well it works as conditions are varied. This information empowers you to optimize performance and maintain robot excellence when conditions change.

If time permits, ask an external reviewer to perform a focused Conceptual Design review on your launcher concept selection and data supporting your selection. Review this critical decision before you move on. Catch any errors in your thinking now so that you do not proceed with a less-than-excellent design.

Defining Robot Details

Now that you have determined how to improve components in your robot, you need to define details that will make your design changes excellent: attractive, elegant, durable, and maybe innovative. This requires attention to details planned for building the finished product. Below are some items to give attention:

- Use quality materials: not warped, nice surface finish, attractive
- Make quality cuts: drill and cut accurately, produce smooth edges
- Make quality joints: solid connections, snug hinges, precisely positioned holes
- Select proper members: suitable cross-sections, adequate strength
- Size motors and transmissions: match motor torque and speed to loads
- Control motion: sensor feedback, stable program controls, user controls
- Buy quality parts: durable, functional, reputable

Consider using commercially available parts when smooth, precise movements are required. Figure 7.2 shows linear actuators that can add quality to a robot system needing linear motion, such as described earlier for moving rings into a launcher. The screw mechanism on the left moves a slide in and out as a motor or servo turns the screw. The rack and pinion actuator on the right slides the rack (toothed bar) linearly as the pinion (gear) is turned by a motor or servo. Commercial products can add excellence in appearance and function compared to products made with limited skill and tools. Be careful to follow game rules that might limit what purchased components are allowed in your robot.

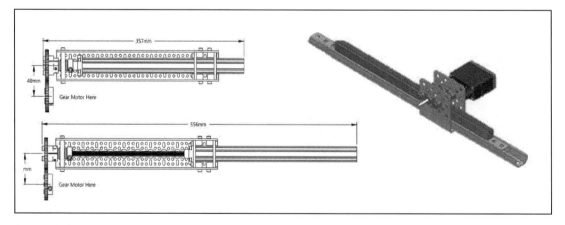

FIGURE 7.2. EXAMPLES OF COMMERCIALLY AVAILABLE LINEAR ACTUATORS

Use CAD to define exact shape and positioning of parts so they fit and move as intended. Vendors often provide CAD files for their parts so you can integrate them into your designs. Developing a full CAD representation of a robot system (or the entire robot) will help you avoid many assembly problems.

Select structural members by the loads they must carry. Types of structural members, their cross-sections, and load stresses are discussed in chapter 4 of *Pre-Engineering Primer, 2nd Edition*. Knowledge of forces, stresses, and material strength can help you choose structural members.

Design excellence also requires that you properly size gears, chains and sprockets, or belts and pulleys to get desired speeds in your robot's moving parts. Loads on robot parts will then define the torque that must be applied to shafts turning these parts. All of this determines the load and rotational speed a motor must deliver to power these parts. With this information, you can use motor curves (load vs. speed data) to find the best motor for powering that part of your robot. Discussion of these power and selection considerations is found in chapters 4 to 6 of the *Pre-Engineering Primer, 2nd Edition*.

When defining robot details for final assembly, be sure to coordinate with teammates who might be assembling other parts. You need to ensure that different parts and different systems will attach as intended. Check now before you begin assembly.

Assembling Completed Robot

Assembling your robot for the next competition can be gratifying as well as disappointing. You hopefully will have significant improvements to implement, adding an element of excellence to the looks and performance of your robot. However, you also are aware of other improvements you have not yet completed. Be encouraged! You are building excellence and you have a plan to achieve it at the right time.

Because time is short, complete finished installation of your improvements as quickly as possible. This requires careful preparation: have necessary parts and tools ready when workers are available. Where appropriate, have groups work in parallel to get multiple improvements installed quickly. If simultaneous work is not feasible, schedule extra work sessions in the order needed to get the robot fully assembled quickly.

Remember that your robot has both hardware and software. As mechanical systems are being assembled, updated software must also be installed. To the extent possible these improvements should be installed during the same period of time.

An important part of robot assembly is quality control. Your chief engineers should oversee assembly to be sure work is not done shabbily. As parts are completed, they should be checked to ensure they meet quality standards. If team members need training, guide them toward competence in fabrication and assembly.

Evaluating Completed Robot

When you have installed your robot improvements and updated software, test your robot to see that everything works as expected. Because testing takes time, combine drive practice and robot testing whenever possible. Your robot's performance will be affected by drive team capabilities, which will change over time.

Your robot testing should focus on scoring performance in practice matches following your team's game strategy. In this way, time is spent practicing the game strategy, which trains the drive team, while gathering data in a relatively authentic testing environment. This makes data relevant for anticipating the upcoming competition.

Prepare your data sheets to collect scoring information in line with your game strategy and solution requirements. This suggests that you record every scoring attempt and success (to calculate scoring probability) and reasons for failed attempts (to identify potential needs for improvement). An example data sheet is shown in Table 7.8.

TABLE 7.8. SAMPLE SCORING SHEET FOR EVALUATION OF ROBOT

Scoring Method	Attempts	Successes	Types of Failures
Auto: WG target zone			__ wrong zone, __ missed zone
Auto: Rings into high goal			__ low, __ high, __ left, __ right, __ jam
Auto: Rings hit power shot			__ low, __ high, __ left, __ right, __ jam
Auto: Park launch line			__ stall, __ missed
Driver: Rings in high goal			___ low, ___ high, ___ left, ___ right, ___ jam
End: WG drop zone			__ missed retrieval, __ time out, __ not over wall
End: Rings into high goal			__ low, __ high, __ left, __ right, __ jam
End: Rings hit power shot			__ low, __ high, __ left, __ right, __ jam

This data sheet records (by tic marks) every scoring attempt, every success, and the number of times different types of failure occur in attempts. From this data, you can calculate probability of success: **Probability = Successes / Attempts**. This information is useful for predicting scores and revising your strategy. Tabulating types of failure helps you identify places where improvements are needed. These kinds of information will enable you to describe your performance to teammates and to other teams, coaches, and judges. They also alert you to types of failures that might occur in a match, some of which you can prevent if you are ready.

Iteration and Pursuit of Excellence

Design iteration is vital to achieving robot excellence. Your challenge is determining when to turn back in your design and how far back to go. Whenever you find that your robot has a problem or new information reveals a better way, you ought to consider iteration.

You usually have several options for iteration. Should you make minor adjustments to your current design or throw out the entire concept and try another? Which current option is best? Make your iteration choice based on considerations of costs and benefits associated with each possible iteration path.

Table 7.9 shows an example cost-benefit tabulation regarding iteration options for a current mechanism that transfers rings from the collection system to the launcher. The current transfer mechanism periodically jams and prevents rings from moving into the launcher. Iteration options are listed in column 1. Column 2 shows an estimate of the likelihood that the iteration can be carried out successfully. Column 3 shows an estimate of the number of person-hours required for the iteration. Column 4 shows other costs that might include monetary costs, conflicts with other parts of the robot, demand on a CNC machine, etc. Column 5 estimates the number of additional points that might be scored in a match if the iteration is implemented. Column 6 lists other benefits from the successful iteration, which might include durability of a mechanism, impacts on your team's reputation, self-image, judging success, etc.

TABLE 7.9. RISK ANALYSIS FOR DESIGN ITERATION OPTIONS

Transfer Redesign Options	Percent Success	Person-Hours	Other Costs	Scoring Benefits	Other Benefits
Operate as currently designed	0%	0	$0	0 pts	None
Operator training, no redesign	10%	3	$0	5 pts	Less damage
Replace bands with flat belts	30%	3	$15	15 pts	Less frustration
New "screw" transfer concept	60%	50	$75	30 pts	Innovation, satisfaction

When your team has discussed the costs and benefits of proposed design iterations, choose the best one, assign work to teammates, allocate resources, and press on to a successful design improvement. Design excellence requires a team to iterate frequently, but also wisely.

Pursuit of robot excellence also requires that you review what you define as excellence, your current level of excellence, and your efforts to improve what matters most. Pause briefly after each competition to revisit your plans for pursuing excellence.

Shortly after a competition:

1. Review the definition of excellence your team hopes to achieve—robot performance, robot design features, your ability to explain design excellence, and your ability to demonstrate excellence.

2. Review how your team and robot demonstrated excellence to other teams, coaches, and judges, at the recent competition.

3. Identify your priorities for making improvements to your robot and team before the next competition.

4. Establish a plan and assign work for making specific improvements in the time available.

After each competition, repeat the process outlined in this chapter to keep a focus on robot excellence. This laser focus will move you closer and closer to excellence as the season progresses.

Listen, for I will speak of excellent things and right things.

Chapter 8: Communicating Your Pursuit for Excellence

A wise person has said, "If it isn't recorded, it didn't happen." This is indeed true when FTC teams are being judged for awards. If you have not effectively recorded your robot design journey, judges do not know how you went from start to finish. You might be able to tell them verbally, but if you have not first written the story, you likely will struggle to explain your robot design journey.

How do you best present your robot design journey? What information should be included and how should it be presented? Perhaps the requirement of a Portfolio for FTC judging is the perfect motivator for learning how to tell your robot development story. Space limitations cause you to be concise. Judges' time constraints require that your portfolio be easy to follow, attractive, and impactful. Judges looking for teams that distinguish themselves look for materials that reflect quality, understanding, and professionalism.

Effective presentations frequently have an opening, a body, and a closing. The opening draws audience attention to the message and sets a context for what follows. The body provides details and supporting information that give credibility to the message. The conclusion drives the audience toward desired understanding, emotions, or action.

The following pages address these three sections of a portfolio by showing examples of content for each section. These few pages do not explain the entire journey, but they might give you ideas for telling a richer story of your robot development journey.

Robot Journey Introduction

What are our robot development goals?

Develop a robot that favorably represents our team, FIRST® and God
1. Competitive robot for in-person and remote competitions
2. Incorporates modularity and industrial design in robot
3. Design decisions supported by analysis and testing
4. Clean, consistent, and concise autonomous and driver-controlled functions
5. Design process that brings significant improvement over the season

What is our underlying team philosophy?

Developing a robot for competitions is only part of our mission. We aspire to learn to work together as Gracious Professionals who bring honor to the team, FIRST®, and God.

Our robot development plan included developing two robots.

Our team chose to develop two different robots over the season. Our first robot was designed for remote matches, where only one robot was on the field. The second robot was designed for more sophisticated performance under in-person competitions.

Developing two robots offered several advantages. A second robot can incorporate significant design changes, as were justified by remote and in-person possibilities. Two robots under development also gave more team members opportunities to lead or make significant contributions. Having two operational robots for later competitions also gives opportunities to practice matches with two robots on the field.

The robot development timeline is shown below. Robot #1 began development after the game release in September. After meet #1 was completed in November, a redesign effort began to create an improved robot for meet #3. Due to delays from COVID-19, the target date for robot #2 was delayed until just prior to meet #4.

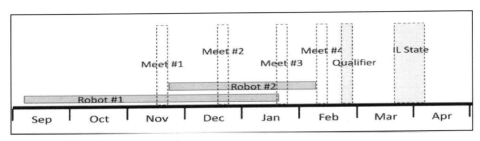

FIGURE 8.1. TIMELINE FOR ROBOT DEVELOPMENT

Robot Journey Body

Our design process included several steps.

Robot design begins by defining game strategy, which determines what the robot must do (user needs). Next the team generates ideas and adapts others' ideas for possible systems in the robot. Prototyping and modeling are used to test ideas. The best ideas are combined into an overall concept for the robot. This concept is built, tested, and refined as necessary to produce a workable robot. At any point in this process, the team may choose to repeat earlier steps (iterate) to improve the needs definition, ideas, concepts, or the final solution. Some systems of the robot may require more iteration than others when more difficult problems are faced.

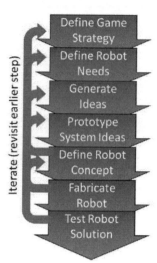

We used design reviews to improve our design.

Design reviews are used to obtain expert help at critical stages of robot development. Identifying problems early saves time and cost. Our team held design reviews of our strategy, solution concept, and computer programs for robot #1 and robot #2.

We made design decisions based upon important criteria.

Design decisions are based on which options best meet stated needs to accomplish our game strategy. A decision matrix (as shown) is used to score each option against criteria for that system. Criteria are weighted 1 to 3 based on importance. Each option is scored 1, 3 or 9 for its fit to each criterion. Total scores are weighted sums of scores for the option, with the highest total indicating the best option.

DESIGN OPTION	Durability	Speed	Accuracy	Complexity	Aesthetics	TOTAL
Weight	2	3	3	2	1	
Belt conveyor	1	3	3	3	1	27
Bucket conveyor	3	3	9	3	1	49
Paddle wheel	3	3	9	3	3	51
Dual screw	9	9	3	9	9	81

Criteria for Selection

Design decisions are based on research on past robots, prototype testing, engineering analysis, and CAD drawings. Ideally, one option is found best and advanced to the next stage of design. When more than one option looks promising, they may be carried forward until additional testing or research shows which is best. If through a design review or additional robot development a serious weakness is found, the team might reconsider earlier options and re-evaluate them with better information to guide the design decision. Yes, design is iterative!

Robot development involved the whole team, experts, and engineering tools.

The following set of pictures shows selected points in the process of developing two robots. The journey begins with team definition of game strategy and defining needs. CAD and design reviews are important tools for design success. We tested prototyped components of the robots and the assembled robots to see how well they met requirements. We iterated many times to improve the robots.

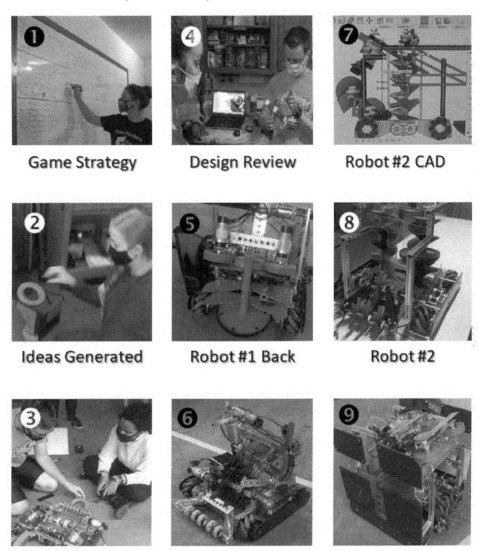

Our team used engineering methods and analysis in the development process.

Ring trajectories were calculated to determine which launch velocities and angles would deliver the ring to the height of the high goal at a distance 2 meters from the goal. Ideally the ring path would be horizontal at entry to the goal.

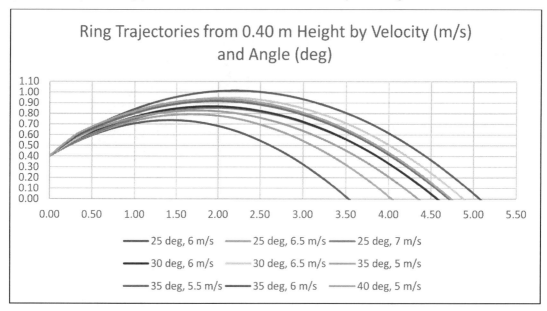

Multiple drawings of rings and collection wheels were used to position collection wheels for smooth collection. A series of drawings modeled ring and wheel interactions as the ring approached the inclined belt conveyor taking rings to the launcher.

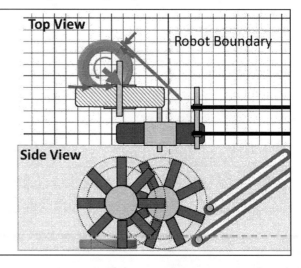

Front paddle wheel
1. Front bristles catch edge of ring
   a. Push ring toward fence but parallel to robot conveyor axis
   b. Fence resists ring advance, applies normal and friction force
   c. Bristle forces and fence forces turn ring counterclockwise
2. Ring follows fence
   a. Ring slides along fence
   b. Ring rotates ccw, so it rolls against fence

Robot Journey Closing

Our robot journey has been an incredible learning experience and adventure. Below we summarize impacts of this journey.

1. Every team member engaged in the robot design process, fabrication, and testing. Four of the 10 members used our CNC machine, 5 used CAD to model parts, and 5 were programmers. Five have learned and used graphic design tools. All have grown as competent speakers. Everyone understands the challenges of design, need for teamwork, and importance of quality fabrication.

2. Our team completed two robots, one for early season use and one for later competitions. Our initial plan was to complete the second robot for our third competition, but it was not ready until the fourth. We learned to use risk analysis to make tough decisions on how to move ahead.

3. We selected our drive team using a test on game rules and actual driving tests. We defined a backup plan for drive team, which was required when our operator became sick and was unable to complete the season.

4. We learned the importance of involving outside people in our robot development. We held five design reviews at important stages of design. This led to our most important innovation, a dual-screw lift system (suggested by a reviewer) for moving rings to the launcher.

5. We learned the importance of drive practice before every competition. When we had no time for drive practice, our robot performed poorly in the competition. When we stopped development earlier and forced drive practice, we were able to find and fix robot problems and gain driver experience that led to better performances.

6. FIRST® is more than robots! Our robot development journey has prepared every member of our team for next steps in life. We are better at time management, teamwork, and making reasoned decisions. Graduating seniors are pursuing varied paths in college, including computer science.

Final Thoughts

Before you leave your robot design experience, reflect on it. Don't forget that designing robots, building robots, fixing robots, and competing with robots is much more than robots. You have learned and applied teamwork while doing demanding work. You have built new relationships. You have learned and practiced professionalism and shown grace to teammates and others. You have learned to think like an engineer, applying creativity and judgment in the face of many challenges. These new life skills will benefit you wherever you go.

Take time to celebrate your achievements in learning, new relationships, and robot development. Share your experiences with others to encourage them to get involved in robot design. Don't underestimate the value of being a role model for others to follow.

Communicate thankfulness. Thank those who have assisted you in your pursuit of excellence. And thank God who gives each of us abilities to learn and do productive work.

Made in the USA
Columbia, SC
29 February 2024